Developer's Guide to Microsoft Enterprise Library, 2nd Edition

Developer's Guide to Microsoft Enterprise Library, 2nd Edition

Dominic Betts
Julián Domínguez
Alex Homer
Hernan de Lahitte
Grigori Melnik
Fernando Simonazzi
Mani Subramanian

978-1-62114-034-4

Contents

Foreword

Microsoft Enterprise Library has a long and accomplished history of providing architectural guidance, recommended practices, and reusable components (called "application blocks") to help developers, architects, and IT professionals efficiently build and deploy line-of-business (LOB) systems.

I've been following the evolution of Enterprise Library since the early days of v1.0 back in 2005. Over the years, the Enterprise Library has achieved a significant reach, seeing over 4 million downloads and powering countless customer applications, partner offerings and ISV applications, as well as our own products and services. During this time Enterprise Library helped drive proven development patterns and practices into the mainstream, including dependency injection, transient fault handling, on-demand autoscaling, and semantic logging.

Many development teams have embraced and benefited from systematic application block reuse, focusing on their business logic as opposed to the underlying "plumbing." Of course, they still configure and customize the application blocks to fit their contexts. As with previous versions, the latest Enterprise Library release contains a wealth of features to help developers, architects, and other IT professionals improve their productivity by providing pre-written, customizable blocks to help them address common cross-cutting concerns in their applications running on-premises on Windows Server or hosted in Windows Azure.

The standout feature for me in the latest release of Enterprise Library is the new Semantic Logging Application Block which will help you to move towards a more structured approach to logging in your applications and help you gain operational and business insights from that log data later.

The new version of Enterprise Library not only includes new application blocks but also extends the library's integration with other technologies, such as ASP.NET MVC and ASP.NET Web API, and simplifies its learning and usage.

The patterns & practices team takes learnability and usability of the library as seriously as developing the blocks themselves. The second edition of the Developer's Guide is a primer on the core architecture of Enterprise Library and its application blocks. It is fully updated to cover the latest application blocks, including: Data Access, Exception Handling, Logging, Semantic Logging, Policy Injection, Transient Fault Handling, and Validation. The in-depth treatment of the Unity Application Block and Unity Interception is presented in a separate sister guide focused on dependency injection and interception.

The guide uses numerous code samples to showcase various capabilities and to go deeper into each of the blocks. It also provides discussions that relate to the use of Enterprise Library in a variety of scenarios and types of application. Perspectives offered from the personae used—from a software developer to an architect to an IT Pro—contain useful insights and trade-offs.

Keep in mind that this guide is not the only learning resources available to you. It is complemented by a comprehensive set of hands-on labs and a reference implemenation. One learns best by practicing. The patterns & practices team provides plenty of ways for you to do so–from small code snippets, to more involved code samples, all the way to a large sample application which you can slice and dice, and experiment with.

The guide is written for a wide variety of software development practitioners–including those experienced with previous versions of Enterprise Library and those who are new to it.

This guide is an absolute must read for anybody looking to put Enterprise Library into practice. Its clear and pragmatic advice is invaluable.

S. Somasegar

Corporate Vice-President
Microsoft Developer Division

Redmond, Washington
November 7, 2013

Foreword to the 1st Edition

You are holding in your hands a book that will make your life as an enterprise developer a whole lot easier.

It's a guide on Microsoft Enterprise Library and it's meant to *guide* you through how to apply .NET for enterprise development. Enterprise Library, developed by the patterns & practices group, is a collection of reusable components, each addressing a specific cross cutting concern—be it system logging, or data validation, or exception management. Many of these can be taken advantage of easily. These components are architecture agnostic and can be applied in a multitude of different contexts.

The book walks you through functional blocks of the Enterprise Library, which include data access, caching, cryptography, exception handling, logging, security, and validation. It contains a large collection of exercises, tricks and tips.

Developing robust, reusable, and maintainable application requires knowledge of design patterns, software architectures and solid coding skills. We can help you develop those skills with Enterprise Library since it encapsulates proven and recommended practices of developing enterprise applications on the .NET platform. Though this guide does not go into the depth of discussions of architecture and patterns, it provides a solid basis for you to discover and implement these patterns from a reusable set of components. That's why I also encourage you to check out the Enterprise Library source code and read it.

This guide is not meant to be a complete reference on Enterprise Library. For that, you should go to MSDN. Instead, the guide covers most commonly used scenarios and illustrates how Enterprise Library can be applied in implementing those. The powerful message manifesting from the guide is the importance of code reuse. In today's world of complex large software systems, high-quality pluggable components are a must. After all, who can afford to write and then maintain dozens of different frameworks in a system—all to accomplish the same thing? Enterprise Library allows you to take advantage of the proven code complements to manage a wide range of task and leaves you free to concentrate on the core business logic and other "working parts" of your application.

Another important emphasis that the guide makes is on software designs, which are easy to configure, testable and maintainable. Enterprise Library has a flexible configuration subsystem driven from either external config files, or programmatically, or both. Leading by example, Enterprise Library itself is designed in a loosely-coupled manner. It promotes key design principles of the separation of concerns, single responsibility principle, principle of least knowledge and the DRY principle (Don't Repeat Yourself). Having said this, don't expect this particular guide to be a comprehensive reference on design patterns. It is not. It provides just enough to demonstrate how key patterns are used with Enterprise Library. Once you see and understand them, try to extrapolate them to other problems, contexts, scenarios.

The authors succeeded in writing a book that is targeted at both those who are seasoned Enterprise Library developers and who would like to learn about the improvements in the latest version, and those, who are brand new to Enterprise Library. Hopefully, for the first group, it will help orientate you and also get a quick refresher of some of the key concepts. For the second group, the book may lower your learning curve and get you going with Enterprise Library quickly.

Lastly, don't just read this book. It is meant to be a practical tutorial. And learning comes only through practice. Experience Enterprise Library. Build something with it. Apply the concepts learnt in practice. And don't forget to share your experience.

In conclusion, I am excited about both the latest release of Enterprise Library and this book.

Scott Guthrie
Corporate Vice-President
Microsoft .NET Developer Platform
Redmond, Washington

May 18, 2010

Preface

About This Guide

When you casually pick up a book in your local bookstore or select one from the endless collection available on your favorite Web site, you're probably wondering what the book actually covers, what you'll learn from it, whether the content is likely to be interesting and useful, and—of course—whether it is actually any good. We'll have a go at answering the first three of these questions here. The final question is one only you can answer. Of course, we would be pleased to hear your opinion through our community Web site at *http://entlib.codeplex.com/*.

What Does This Guide Cover?

As you can probably tell from the title, this guide concentrates on how you can get started with Enterprise Library. It will help you learn how to use Enterprise Library in your applications to manage your crosscutting concerns, simplify and accelerate your development cycle, and take advantage of proven practices. Enterprise Library is a collection of prewritten code components that have been developed and fine-tuned over many years. You can use them out of the box, modify them as required, and distribute them with your applications. You can even use Enterprise Library as a learning resource. It includes the source code that demonstrates Microsoft .NET programming techniques and the use of common design patterns that can improve the design and maintainability of your applications. By the way, if you are not familiar with the term crosscutting concerns, don't worry; we'll explain it as we go along.

Enterprise Library is an extensive collection, with a great many moving parts. To the beginner, knowing how to best take advantage of it is not completely intuitive. Therefore, in this guide we'll help you to quickly understand what Enterprise Library is, what it contains, how you can select and use just the specific features you require, and how easy it is to get started using them. You will see how you can quickly and simply add Enterprise Library to your applications, configure it to do exactly what you need, and then benefit from the simple-to-use, yet extremely compelling opportunities it provides for writing less code that achieves more.

The first chapter of this guide discusses Enterprise Library in general, and provides details of the individual parts so that you become familiar with the framework as a whole. The aim is for you to understand the basic principles of each of the application blocks in Enterprise Library, and how you can choose exactly which blocks and features you require. Chapter 1 also discusses the fundamentals of using the blocks, such as how to configure them, how to instantiate the components, and how to use these components in your code.

The remaining seven chapters discuss in detail the application blocks that provide the basic crosscutting functionality such as data access, logging, and exception handling. These chapters explain the concepts that drove development of the blocks, the kinds of tasks they can accomplish, and how they help you implement many well-known design patterns. And, of course, they explain—by way of code extracts and sample programs—how you actually use the blocks in your applications. After you've read each chapter, you should be familiar with the block and be able to use it to perform a range of functions quickly and easily, in both new and existing applications.

This guide also includes a section called Tales from the Trenches, which contains case studies from the real world, including experiences of developers using Enterprise Library for the very first time. Additional case studies get posted online as they become available (*http://msdn.com/entlib*), so make sure to check them out. If you want to share your story with the developer community at large, send a draft to *ourstory@microsoft.com* and we'll work with you to get your story across.

Finally, the appendices present more detailed information on specific topics that you don't need to know about in detail to use Enterprise Library, but are useful as additional resources and will help you understand how features such as dependency injection, interception, and encryption fit into the Enterprise Library world.

All of the chapters include references to additional resources such as books, blog posts, and papers that will provide additional detail if you want to explore some of the topics in greater depth. For your convenience, there is a bibliography online that contains all the links so that these resources are just a click away: *http://aka.ms/el6biblio*.

You can also download and work through the Hands-On Labs for Enterprise Library, which are available at *http://aka.ms/el6hols*.

What This Guide Does Not Cover

The aim of this guide is to help you learn how to benefit from the capabilities of Enterprise Library. It does not describe the common design patterns in depth, or attempt to teach you about application architecture in general. Instead, it concentrates on getting you up to speed quickly and with minimum fuss so you can use Enterprise Library to manage your crosscutting concerns.

Enterprise Library is designed to be extensible. You can extend it simply by writing custom plug-in providers, by modifying the core code of the library, or even by creating entirely new blocks. In this guide, we provide pointers to how you can do this and explain the kinds of providers that you may be tempted to create, but it is not a topic that we cover in depth. These topics are discussed more fully in the *Enterprise Library Reference Documentation*, and in the many other resources available from our community Web site at *http://www.codeplex.com/entlib*.

For more information about the Dependency Injection (DI) design pattern and associated patterns, see "*Chapter 2 – Dependency Injection*" in the Unity Developer's Guide.

How Will This Guide Help You?

If you build applications that run on the Microsoft .NET Framework, whether they are enterprise-level business applications or even relatively modest Windows Forms, Windows Presentation Foundation (WPF), Windows Communication Foundation (WCF), or ASP.NET applications, you can benefit from Enterprise Library. This guide helps you to quickly grasp what Enterprise Library can do for you, presents examples that show it in action, and make it easier for you to start experimenting with Enterprise Library.

The sample applications are easy to assimilate, fully commented, and contain code that demonstrates all of the main features. You can copy this code directly into your applications if you wish, or just use it as a guide when you need to implement the common functionality it provides. The samples are console-based applications that contain separate procedures for each function they demonstrate. You can download these samples from *http://go.microsoft.com/fwlink/p/?LinkID=304210*.

Finally, what is perhaps the most important feature of this guide is that it will hopefully allay any fears you may have about using other people's code in your applications. By understanding how to select exactly the features you need, and installing the minimum requirements to implement these features, you will see that what might seem like a huge and complicated framework is actually a really useful set of individual components and features from which you can pick and choose—a candy store for the architect and developer.

Who's Who

The guide includes discussions and examples that relate to the use of Enterprise Library in a variety of scenarios and types of application. A panel of experts provides a commentary throughout the book, offering a range of viewpoints from developers with various levels of skill, an architect, and an IT professional. The following table lists the various experts who appear throughout the guide.

Markus is a software developer who is new to Enterprise Library. He is analytical, detail-oriented, and methodical. He's focused on the task at hand, which is building a great LOB application. He knows that he's the person who's ultimately responsible for the code.

> "I want to get started using Enterprise Library quickly, so I want it to be simple to integrate with my code and easy to configure with plenty of sensible defaults."

Beth is a developer who used Enterprise Library some time ago but abandoned it in her more recent projects. She is interested in re-evaluating it but her primary concern is that it shouldn't be an all-or-nothing deal.

> "I'm happy using libraries and frameworks but I don't want to get tied into dependencies that I don't need. I want to be able to use just the components I need for the task in hand."

Jana is a software architect. She plans the overall structure of an application. Her perspective is both practical and strategic. In other words, she considers not only what technical approaches are needed today, but also what direction a company needs to consider for the future. Jana has worked on many projects that have used Enterprise Library as well as other libraries. Jana is comfortable assembling a best-of-breed solution using multiple libraries and frameworks.

> "It's not easy to balance the needs of the company, the users, the IT organization, the developers, and the technical platforms we rely on while trying to ensure component independence."

Carlos is an experienced software developer and Enterprise Library expert. As a true professional, he is well aware of the common crosscutting concerns that developers face when building line-of-business (LOB) applications for the enterprise. His team is used to rely on Enterprise Library and they are happy to see continuity in Enterprise Library releases. Quality, support, and ease of migration are his primary concerns.

> "Our existing LOB applications use Enterprise Library for crosscutting concerns. This provides a level of uniformity across all our systems that make them easier to support and maintain. We want to be able to migrate our existing applications to the new version with a minimum of effort."

Poe is an IT professional who's an expert in deploying and managing LOB applications. Poe has a keen interest in practical solutions; after all, he's the one who gets paged at 3:00 AM when there's a problem. Poe wants to be able to tweak application configuration without recompiling or even redeploying them in order to troubleshoot.

"I want a consistent approach to configuration for all our applications both on-premises and in the cloud and plenty of flexibility for logging to make it easy to manage and troubleshoot our applications."

What Do You Need to Get Started?

The prerequisites for using this guide are relatively simple. You'll need to be relatively experienced in C#, and understand general object-oriented programming techniques. The system requirements and prerequisites for using Enterprise Library are:

- Supported architectures: x86 and x64.
- Operating system: Microsoft Windows 8.1, Microsoft Windows 7, Windows Server 2008 R2, Windows Server 2012.
- Microsoft .NET Framework 4.5 or 4.5.1.
- For a rich development environment, the following are recommended:
    - Microsoft Visual Studio 2012 or 2013, Professional, Ultimate, or Express editions.
- For the Data Access Application Block, the following is also required:
    - A database server running a database that is supported by a .NET Framework 4.5 data provider.
- For the Logging Application Block, the following are also required:
    - If you are using the Message Queuing (MSMQ) Trace Listener to store log messages, you need the Microsoft Message Queuing (MSMQ) components installed.
    - If you are using the Database Trace Listener to store log messages, you need access to a database server.
    - If you are using the E-mail Trace Listener to store log messages, you need access to an SMTP server.
- For the Semantic Logging Application Block, the following may be required:
    - If you are using the SQL Database Sink to store log messages, you need access to a SQL Server database server.
    - If you are using the Windows Azure Sink to store log messages, you need access to a Windows Azure storage account. You must also install the Windows Azure SDK Version 1.8.

You can use the NuGet package manager in Visual Studio to install the Enterprise Library assemblies that you need in your projects.

Other than that, all you require is some spare time to sit and read, and to play with the example programs. Hopefully you will find the contents interesting (and perhaps even entertaining), as well as a useful source for learning about Enterprise Library.

The Team Who Brought You This Guide

Vision	Grigori Melnik
Authors	Dominic Betts, Julián Domínguez, Alex Homer, Hernan de Lahitte, Grigori Melnik, Fernando Simonazzi, and Mani Subramanian
Foreword by	S. Somasegar
Foreword to the 1st edition by	Scott Guthrie
Technical Reviewers	Fabian Fernandez, Mariano Grande, Hernan de Lahitte, Randy Levy, Vladimir Melnik, and Dan Piessens
Graphic Artist and Book Layout	Chris Burns
Cover Design	Grigori Melnik, Masashi Narumoto, RoAnn Corbisier, Nelly Delgado and Chris Burns
Editor	RoAnn Corbisier
Production	Nelly Delgado

The Enterprise Library 6 Development Team

Product/Program Management	Grigori Melnik (Microsoft Corporation).
Architecture/ Development	Julian Dominguez (Microsoft Corporation), Fernando Simonazzi and Edgardo Rossetto (Clarius Consulting), and Hernan de Lahitte (Digit Factory)
Testing	Mani Subramanian (Microsoft Corporation), Mariano Grande (Digit Factory), Rathi Velusamy, Lavanya Selvaraj, and Shebu Kollam Valappil (Infosys Technologies Ltd.)
Performance/ Testing	Carlos Farre (Microsoft Corporation), Naveen Pitipornvivat and Soranai Security Chantarayotha (Adecco)
Documentation	Dominic Betts (Content Master Ltd), Grigori Melnik, Julian Dominguez and Mani Subramanian (Microsoft Corporation), and Fernando Simonazzi (Clarius Consulting)
Editing/Pre- and Post Production	RoAnn Corbisier, Nelly Delgado, Handan Selamoglu and Jane Sinyagina (Microsoft Corporation), Chris Burns (Linda Werner & Associates)
Release Management	Nelly Delgado and Grigori Melnik (Microsoft Corporation)
Advisory Council	Fabian Fernandez Bargas (TellMi), Joseph M. Blank (Online Business Systems), Bryan Clark (Ethos Solutions), Gabriele Giuseppini (Optiver Holding B.V.), Deon Heyns, Matt Kean (Caterpillar Inc), Ercenk Keresteci and Trent Swanson (Full Scale 180, Inc.), Randy Levy, Jeremy Likness (Wintellect), Andrei Marukovich (AB SCIEX), Ksenia Mukhortova (Intel), Peter Nilsson (Volvo Information Technology AB), Dan Piessens (Centare), Peter Ritchie (Peter Ritchie Inc.), Luke G. Sigler (Qualcomm Life), Jon Wagner (eMoney Advisor), Sebastian Weber, Bill Wilder (DevPartners), Matias Woloski (Auth0), Walter Wu (Royal Bank of Canada), Nuno Centeno, Jason Hogg, Tom Hollander, Robert Jarratt, Michael Lanzetta, Valery Mizonov, Rafael Fernandez Moctezuma, Tyler Ohlsen, Timothy Stockstill, Christopher Tavares, Michael Thomassy, and Rob Vettor (Microsoft Corporation)
Community	Attendees at patterns & practices summits and symposia, //Build, TechReady, and TechEd conferences who provided informal feedback; and Unity users who commented on this guide on CodePlex, through our blogs, surveys and via e-mail. Thank you!

1 Welcome to the Library

Introduction to the Microsoft Enterprise Library

Meet the Librarian

Before we begin our exploration of Microsoft Enterprise Library and the wondrous range of capabilities and opportunities it encompasses, you need to meet the Librarian. In the early days we called him Tom, sometimes we called him Chris, and for the past six years we call him Grigori. He—in collaboration with an advisory board of experts from the industry and other internal Microsoft product groups, and a considerable number of other community contributors—is the producer and guardian of the Microsoft Enterprise Library.

Since its inception as a disparate collection of individual application blocks, the Librarian has guided, prodded, inspired, and encouraged his team to transform it into a comprehensive, powerful, easy-to-use, and proven library of code that can help to minimize design and maintenance pain, maximize development productivity, and reduce costs. And now in version 6, it contains even more built-in goodness that should make your job easier. It's even possible that, with the time and effort you will save, Enterprise Library can reduce your golf handicap, help you master the ski slopes, let you spend more time with your kids, or just make you a better person. However, note that the author, the publisher, and their employees cannot be held responsible if you just end up watching more TV or discovering you actually have a life.

If you've used Enterprise Library before, you'll find version 6 is easy to get started with. There are some changes, some features have been deprecated, and new features have been added, but it continues to address the common cross-cutting concerns that developers face building line-of-business applications both on-premises and in the cloud.

What You Get with Enterprise Library

Enterprise Library is made up of a series of application blocks, each aimed at managing specific *crosscutting concerns*. In case this concept is unfamiliar, crosscutting concerns are those annoying tasks that you need to accomplish in several places in your application. When trying to manage crosscutting concerns there is often the risk that you will implement slightly different solutions for each task at each location in your application, or that you will just forget them altogether. Writing entries to a system log file or Windows Event Log, and validating user input are typical crosscutting concerns. While there are several approaches to managing them, the Enterprise Library application blocks make it a whole lot easier by providing generic and configurable functionality that you can centralize and manage.

What are *application blocks*? The definition we use is "pluggable and reusable software components designed to assist developers with common enterprise development challenges." Application blocks help address the kinds of problems developers commonly face from one line-of-business project to the next. Their design encapsulates the Microsoft recommended practices for Microsoft .NET Framework-based applications, and developers can add them to .NET-based applications and configure them quickly and easily.

As well as the application blocks, Enterprise Library contains an optional configuration tool, plus a set of core functions that manage tasks applicable to all of the blocks. Some of these functions—routines for handling configuration and serialization, for example—are exposed and available for you to use in your own applications.

And, on the grounds that you need to learn how to use any new tool that is more complicated than a hammer or screwdriver, Enterprise Library includes a range of sample applications, Quickstarts, descriptions of key scenarios for each block, hands-on labs, and comprehensive reference documentation. You even get all of the source code and the unit tests that the team created when building each block (the team follows a test-driven design approach by writing tests before writing code). So you can understand how it works, see how the team followed good practices to create it, and then modify it if you want it to do something different. Figure 1 shows the big picture for Enterprise Library.

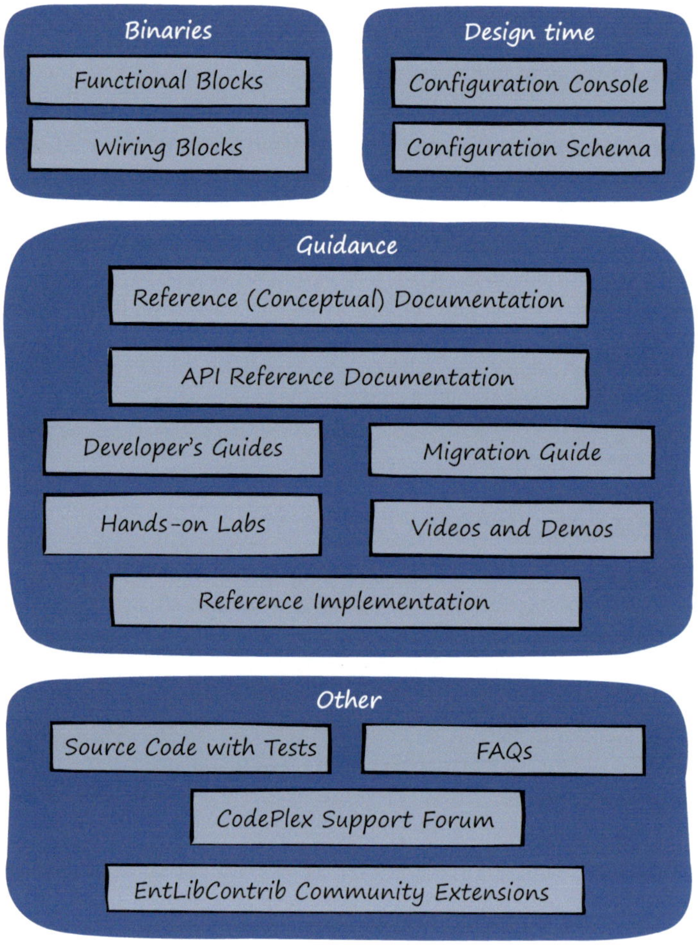

FIGURE 1
Enterprise Library—the big picture

Things You Can Do with Enterprise Library

If you look at the documentation, you'll see that Enterprise Library today actually contains eight application blocks. However, there are actually only six blocks that "do stuff"—these are referred to as *functional* blocks. The other two are concerned with "wiring up stuff" (the *wiring* blocks). What this really means is that there are six blocks that target specific crosscutting concerns such as logging, data access, and validation. The other two, the Unity Dependency Injection mechanism and the Policy Injection Application Block, are designed to help you implement more loosely coupled, testable, and maintainable systems. There's also some shared core pieces used in all the blocks. This is shown in Figure 2.

With this release of Enterprise Library, there are fewer dependencies between the blocks. You can choose to use just the blocks that are relevant to your application.

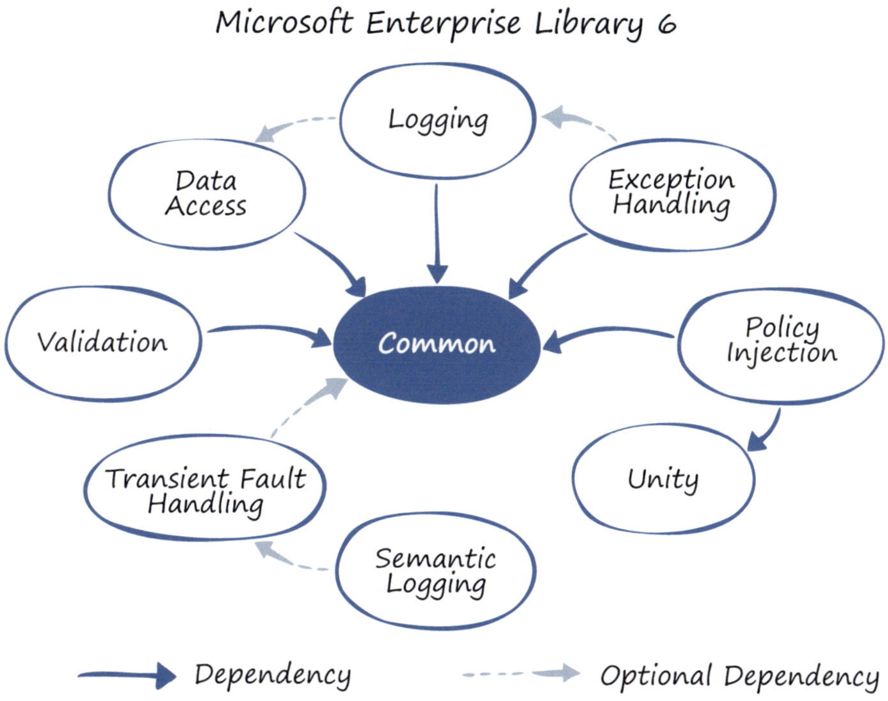

Figure 2
The parts of Enterprise Library

In this book we'll be concentrating on the six functional blocks. If you want to know more about how you can use Unity and the Policy Injection Application Block, check out the *Dependency Injection with Unity* guide. It describes the capabilities of Unity as a dependency injection mechanism and the use of policy injection in more detail.

You can use Unity to add dependency injection to your applications and to use interception techniques to address additional crosscutting concerns specific to your applications.

The following list describes the crosscutting scenarios you'll learn about in this book:

- **Data Access**. The Data Access Application Block simplifies many common data access tasks such as reading data for display, passing data through application layers, and submitting changed data back to the database system. It includes support for both stored procedures and in-line SQL, can expose the data as a sequence of objects for client-side querying, and provides access to the most frequently used features of ADO.NET in simple-to-use classes.

- **Exception Handling**. The Exception Handling Application Block lets you quickly and easily design and implement a consistent strategy for managing exceptions that occur in various architectural layers of your application. It can log exception information, hide sensitive information by replacing the original exception with another exception, and maintain contextual information for an exception by wrapping the original exception inside another exception.
- **Transient Fault Handling**. The Transient Fault Handling Application Block makes your application more robust by providing the logic for handling transient faults. It does this in two ways. First, the block includes logic to identify transient faults for a number of common cloud-based services in the form of detection strategies. Second, the application block enables you to define your retry strategies so that you can follow a consistent approach to handling transient faults in your applications. The block also helps you to perform retries if you are dealing with asynchronous, task-based code.
- **Logging**. The Logging Application Block simplifies the implementation of common logging functions such as writing information to the Windows Event Log, an e-mail message, a database, Windows Message Queuing, a text file, or a custom location.
- **Semantic Logging**. The Semantic Logging Application Block enables you to use the **EventSouce** class to write strongly typed log messages from your application. This enables you to write log messages with a consistent structure and format and to collect and process log messages out-of-process.
- **Validation**. The Validation Application Block provides a range of features for implementing structured and easy-to-maintain validation mechanisms using attributes and rule sets, and integrating with most types of application interface technologies.

If you have used previous versions of Enterprise Library and are wondering what happened to the Caching Application Block, Security Application Block and Cryptography Application Block as well as some other functionality, the answer is that these have been deprecated. Many scenarios supported by these blocks are now better supported by the .NET platform. Our deprecation philosophy is outlined in *this post* by the Librarian. The V6 project was focused on ensuring Enterprise Library's close alignment to the current platform (.NET framework 4.5) with a goal of reducing Enterprise Library's footprint by leveraging platform capabilities and improvements. For more details, see the *Migration Guide*.

Why You Should Use Enterprise Library

As you can see from the previous section, Enterprise Library provides a comprehensive set of features that can help you to manage your crosscutting concerns though a reusable set of components and core functionality. Of course, like many developers, you may suffer from the well-known NIH (not invented here) syndrome. But, seriously, isn't it about time that every developer on your team stopped writing his or her own logging framework or other "plumbing"? It's a commonly accepted fact that the use of standard and proven code libraries and components can save development time, minimize costs, reduce the use of precious test resources, and decrease the overall maintenance effort. In the words of the Librarian, "These days you cannot afford not to reuse."

You can download the Nucleus Research 2009 Report on Microsoft patterns & practices, which reviews the key components, benefits, and includes direct feedback from software architects and developers who have adopted patterns & practices deliverables in their projects and products from *http://www.microsoft.com/en-us/download/confirmation.aspx?id=11522* .

And it's not as though Enterprise Library is some new kid on the block that might morph into something completely different next month. Enterprise Library as a concept has been around for many years, and has passed through six full releases of the library as well as intermediate incremental releases.

Enterprise Library continues to evolve along with the capabilities of the .NET Framework. As the .NET Framework has changed over time, some features that were part of Enterprise Library were subsumed into the core, while Enterprise Library changed to take advantage of the new features available in both the .NET Framework and the underlying system. Examples include new programming language capabilities, and the use of asynchronous techniques, and the Task Parallel Library. You can also use Enterprise Library in your Windows Azure cloud-based applications as well as in your on-premises applications. Yet, even in version 6, the vast majority of the code is entirely backwards compatible with applications written to use Enterprise Library 2.0.

Upgrading to a new version of Enterprise Library may require some changes to your applications, but the guidance that accompanies Enterprise Library will help you to identify and make the necessary changes.

You can also use Enterprise Library as learning material—not only to implement design patterns in your application, but also to learn how the development team applies patterns when writing code. Enterprise Library embodies many design patterns, and demonstrates good architectural and coding techniques. The source code for the entire library together with unit tests is provided, so you can explore the implementations and reuse the techniques in your own applications.

And, finally, it is free! Or rather, it is distributed under the Microsoft Public License (MS-PL) that grants you a royalty-free license to build derivative works, and distribute them free—or even sell them. You must retain the attribution headers in the source files, but you can modify the code and include your own custom extensions. Do you really need any other reasons to try Enterprise Library?

The Enterprise Library source code and tests are great learning resources, not just for Enterprise Library but for .NET development as well.

You'll notice that, even though we didn't print *"Don't Panic!"* in large friendly letters on the cover, this book does take a little time to settle down into a more typical style of documentation, and start providing practical examples. However, you can be sure that—from here on in—you'll find a whole range of guidance and examples that will help you master Enterprise Library quickly and easily. There are resources to help if you're getting started with Enterprise Library, and there's help for existing users as well (such as the breaking changes and migration information for previous versions) available on *CodePlex*. You can also visit the Preview section of the site to see what the Enterprise Library team is working on as you read this guide.

Some Fundamentals of Enterprise Library

Before we dive into our tour of the application blocks and features of Enterprise Library, you need to grasp some fundamentals. In this chapter, the Librarian will help you explore topics such as how to install and deploy the library, and how to perform initial configuration. After that, you'll be free to skip to any of the other chapters and learn more about the ways that each block helps you to simplify your code and manage your crosscutting concerns. For more information about the topics covered in this chapter, see the *Enterprise Library Reference Documentation*.

Choosing Which Blocks to Install

NuGet makes it very easy to get started. Installing a specific application block package downloads all the required assemblies and adds all the required references in one easy step.

Enterprise Library is a "pick and mix" candy store, where you choose just the features you want to use and simply disregard the rest. Once you have chosen which Enterprise Library blocks to use in your application, you need to add the appropriate assemblies to your project: for the Enterprise Library blocks, you can use the NuGet package manager in Visual Studio to handle this for you.

A NuGet package typically contains one or more assemblies, links to other NuGet packages that the current one depends on, and some configuration settings for your project. In some cases, NuGet packages also include additional project resources such as XML schema files or readme files.

From the perspective of Enterprise Library, the great advantage of NuGet is that it automatically adds everything that you need to use a block (including any dependencies) to your project in one, easy step. If you've not used NuGet before, you can find out more at "*Welcome to NuGet Docs*."

When NuGet installs a package, it places all the assemblies that make up a package and its dependencies in a folder within your solution. Therefore NuGet doesn't install or change anything on your machine, it just modifies the project. Because the correct version of all of the Enterprise Library assemblies that your project uses are now part of the project, you'll find that it's much easier to deploy your application with all the correct dependencies.

In some cases, an Enterprise Library block consists of more than one NuGet package. This happens when you don't necessarily need all of the features offered by a block. For example, you could install just the EnterpriseLibrary.ExceptionHandling NuGet package. However, if you want to log certain types of exception, you can also install the EnterpriseLibrary.ExceptionHandling.Logging NuGet package. Not surprisingly, if you begin by trying to install the EnterpriseLibrary.ExceptionHandling.Logging package, NuGet will automatically install both the EnterpriseLibrary.ExceptionHandling and EnterpriseLibrary.Logging packages that it depends on.

The configuration tool will automatically add the required block configuration to your application configuration file with the default configuration when required. For example, when you add a Logging handler to an Exception Handling block policy, the configuration tool will add the Logging block to the configuration with the default settings.

Installing Enterprise Library

There is no Enterprise Library installation; you add the blocks you need to any Visual Studio project by using the NuGet package manager. You can find all the Enterprise Library blocks in the **Manage NuGet Packages** dialog in Visual Studio by searching online for **EnterpriseLibrary**. You can also use the **Package Manager Console** in Visual Studio if you prefer to work on the command line.

> *NuGet enables you to add packages to projects and solutions. If you add a package to a solution, you can then use NuGet to add references to the package to individual projects within the solution.*

If you want to examine the source code, and perhaps even modify it to suit your own requirements, you can download the *EnterpriseLibrary 6-source.exe* (a self-extractable zip file). You can also download the binaries, Reference Implementation, and QuickStarts from the *Microsoft Download Center*.

In NuGet, all new Enterprise Library packages are tagged with the '**entlib6**' keyword. This makes it easy to search for the latest packages.

The zip file also contains a folder called **scripts** that includes batch files to install database files and other features. There are also batch files that you can use to compile the entire library source code, and to copy all the assemblies to the **bin** folder within the source code folders, if you want to rebuild the library from the source code.

Assemblies and References

It's not uncommon, when people first look at Enterprise Library, to see a look of mild alarm spread across their faces. Yes, there are quite a few assemblies, but remember:

- You only need to use those directly connected with your own scenario.
- Several are required for only very special situations.
- The runtime assemblies you will use in your applications are mostly less than 100 KB in size; and the largest of all is only around 500 KB.
- In most applications, the total size of all the assemblies you will use will be between 1 and 2 MB.
- NuGet will make sure that you have all the required assemblies for the blocks that you are using.

GAC or Bin, Signed or Unsigned?

All of the assemblies are provided as precompiled signed versions that NuGet places in a folder within your project. This helps to ensure that your project references the correct version of the assemblies you are using. However, you can install the assemblies into the global assembly cache (GAC) if you wish.

NuGet can install a package to either a Visual Studio solution or project. NuGet never makes any changes outside of a solution. For example, NuGet never installs assemblies into the GAC.

NuGet adds references in your project to the compiled assemblies it downloaded, these assemblies are automatically copied to the bin folder when you build your solution. This approach gives you simple portability and easy installation.

Alternatively, you can install the source code for Enterprise Library and use the scripts provided to compile unsigned versions of the assemblies. This is useful if you decide to modify the source code to suit your own specific requirements. You can strong name and sign the assemblies using your own credentials if required.

For more information about side-by-side operation and other deployment issues, see the *Enterprise Library Reference Documentation*.

Importing Namespaces

After you reference the appropriate assemblies in your projects, you will probably want to add **using** statements to your project files to simplify your code and avoid specifying types using the full namespace names.

You will also need to import the namespaces for the specific application blocks you are using. Most of the Enterprise Library assemblies contain several namespaces to organize the contents. For example, the Semantic Logging Application Block includes the following namespaces.

- Microsoft.Practices.EnterpriseLibrary.SemanticLogging
- Microsoft.Practices.EnterpriseLibrary.SemanticLogging.Database
- Microsoft.Practices.EnterpriseLibrary.SemanticLogging.Etw
- Microsoft.Practices.EnterpriseLibrary.SemanticLogging.Etw.WindowsService
- Microsoft.Practices.EnterpriseLibrary.SemanticLogging.WindowsAzure

Configuring Enterprise Library

Enterprise Library offers users several options for configuring the various blocks. Typically, you use an extremely flexible programmatic approach to configure the blocks: this is the approach used in the examples in this guide. If you have used a previous release of Enterprise Library, you will have used either a declarative approach based on XML configuration files and the Configuration Tool, or the fluent configuration API. For more information about these legacy approaches, you should read the reference documentation and the *Developer's Guide for Enterprise Library 5.0*.

In previous version of Enterprise Library, the preferred configuration approach was declarative, using the Configuration Tool to edit the configuration files.

Diving in with an Example

To demonstrate the configuration features of Enterprise Library, we provide a sample application that you can download and run on your own computer. You can run the executable directly from the bin\Debug folder, or you can open the solution named **Configuration** in Microsoft Visual Studio to see the code and run it under Visual Studio.

> *Depending on the version of the operating system you are using, you may need to execute the application under the context of an account with administrative privileges. If you are running the sample from within Visual Studio, start Visual Studio by right-clicking the entry in your **Start** menu and selecting **Run as administrator**.*

One point to note about the sample application is that it creates a folder named **Temp** in the root of your C: drive if one does not already exist, and writes the text log files there so that you can easily find and view them.

Configuration Classes

For most blocks, you can simply create the objects you need. However, the Logging Application Block includes a special configuration **LoggingConfiguration** class. Typically, you instantiate any necessary supporting objects before you create the configuration class, and then initialize the application block object by passing it the configuration object. The following code illustrates how to create a **LoggingConfiguration** object and then initialize a **LogWriter** object. The **LogWriter** class is part of the Logging Application Block.

```
// Create filters
PriorityFilter priorityFilter = new PriorityFilter(...);
LogEnabledFilter logEnabledFilter = new LogEnabledFilter(...);
CategoryFilter categoryFilter = new CategoryFilter(...);

// Create trace listeners
FlatFileTraceListener flatFileTraceListener =
                      new FlatFileTraceListener(...);

// Build Configuration
LoggingConfiguration config = new LoggingConfiguration();
config.Filters.Add(priorityFilter);
config.Filters.Add(logEnabledFilter);
config.Filters.Add(categoryFilter);
config.AddLogSource("General", SourceLevels.All, true, flatFileTraceListener);

// Configure the LogWriter instance
LogWriter defaultWriter = new LogWriter(config);
// Create filters
PriorityFilter priorityFilter = new PriorityFilter(...);
LogEnabledFilter logEnabledFilter = new LogEnabledFilter(...);
CategoryFilter categoryFilter = new CategoryFilter(...);
```

The code shown in this example is adapted slightly to make it easier to read from the code in the method **BuildProgrammaticConfig** in the sample application.

For other blocks, you can simply instantiate the required objects and start using them. For example, in the Validation Application Block, you can create validators directly in code as shown in the following example.

```
Validator[] valArray = new Validator[] {
                      new NotNullValidator(true, "Value can be NULL."),
                      new StringLengthValidator(
                          5, RangeBoundaryType.Inclusive,
                          5, RangeBoundaryType.Inclusive,
                          "Value must be between {3} ({4}) and {5} ({6}) chars.")
};

Validator orValidator = new OrCompositeValidator(
                          "Value can be NULL or a string of 5 characters.", valArray);

// This will not cause a validation error.
orValidator.Validate(null, valResults);

// This will cause a validation error.
orValidator.Validate("MoreThan5Chars", valResults);
```

Instantiating and Using Enterprise Library Objects

After you have referenced the assemblies you need, imported the required namespaces, and configured your application, you can start to think about creating instances of the Enterprise Library objects you want to use in your applications. As you will see in each of the following chapters, the Enterprise Library application blocks are optimized for use as loosely coupled components in almost any type of application. Typically, if you are using declarative configuration, you will use a factory provided by the block to create and configure the objects that you need in your application.

Enterprise Library Objects and Factories

Each of the application blocks in Enterprise Library contains one or more core objects that you typically use to access the functionality of that block. An example is the Exception Handling Application Block, which provides a class named **ExceptionManager** that exposes the methods you use to pass exceptions to the block for handling. The following table lists the commonly used objects for each block.

Functional Application Block	Non-static Instance or Factory	Static Instances or Factories
Data Access	Database GenericDatabase SqlDatabase SqlCeDatabase OracleDatabase DatabaseProviderFactory	DatabaseFactory
Exception Handling	ExceptionManager	ExceptionPolicyFactory ExceptionPolicy
Transient Fault Handling	RetryManager RetryPolicyFactory	
Logging	LogWriter LogEntry TraceManager LogWriterFactory	Logger
Semantic Logging	ObservableEventListener ConsoleSink FlatFileSink RollingFlatFileSink SqlDatabaseSink WindowsAzureTableSink	
Validation	ValidationFactory ConfigurationValidatorFactory AttributeValidatorFactory ValidationAttributeValidatorFactory	ValidatorFactory

This table includes the task-specific objects in some blocks that you can create directly in your code in the traditional way using the ***new*** *operator. For example, you can create individual validators from the Validation Application Block, or log entries from the Logging Application Block. We show how to do this in the examples for each application block chapter.*

To use the features of an application block, all you need to do is create an instance of the appropriate object, facade, or factory listed in the table above and then call its methods. The behavior of the block is controlled by the configuration you specified, and often you can carry out tasks such as exception handling and logging with just a single line of code. Even tasks such as accessing data or validating instances of your custom types require only a few lines of simple code. So, let's look at how you create instances of the Enterprise Library objects you want to use.

Creating Instances of Enterprise Library Types

In this release of Enterprise Library, the recommended approach to creating instances of Enterprise Library objects is to use the programmatic approach and instantiate the objects directly. You may decide to store some configuration information externally (such as in a custom configuration section or in the Windows Azure service settings) and use this information when you create the Enterprise Library objects programmatically. In this way, you can expose just those settings that you want to make available, rather than all the settings, which is the case when you use declarative configuration. Each block is self-contained and does not have dependencies on other blocks for basic operations. Typically, creating an instance is a simple operation that takes only a few lines of code.

You could use Unity, or another dependency injection container, to manage your Enterprise Library objects and their dependencies with the corresponding lifecycles. Unlike in the previous release, it is now your responsibility to register and resolve the types you plan to use. Unity 3 now supports the registration by convention to make it easier to do so. See the *Dependency Injection with Unity* guide for more info.

The chapters that cover the individual blocks provide the details of how to create the objects relevant to that block. For now, you'll see an example from the Data Access Application Block that illustrates the core principles. Typically, you start by creating a factory object and then use that factory to construct an instance of the required type.

The following code sample first creates a **DatabaseProviderFactory** instance and then use two different methods to create **Database** objects.

```
DatabaseProviderFactory factory = new DatabaseProviderFactory();

Database defaultDB = factory.CreateDefault();

Database namedDB = factory.Create("ExampleDatabase");
```

The **DatabaseProviderFactory** class provides several overloaded constructors that enable you to specify from where the factory should obtain the configuration data that it needs. In the case of the Data Access Application Block, the key configuration data is the connection strings for the databases you are using in your application.

The Example Applications

To help you understand how you can use Enterprise Library and each of the seven application blocks covered in this guide, we provide a series of simple example applications that you can run and examine. Each is a console-based application and, in most cases, all of the relevant code that uses Enterprise Library is found within a series of routines in the Program.cs file. This makes it easy to see how the different blocks work, and what you can achieve with each one.

The examples use the simplest approach for creating the Enterprise Library objects they require (in most cases using a factory class or instantiating the required objects directly), most define the configuration information programmatically but for reference some also contain equivalent declarative configuration in their configuration files. Each of the options in the examples exercises specific features of the relevant block and displays the results. You can open the solutions for these examples in Visual Studio, or just run the executable file in the bin\debug folder and view the source files in a text editor if you prefer.

To obtain the example applications, go to *http://go.microsoft.com/fwlink/p/?LinkID=304210*.

Summary

This brief introduction to Enterprise Library will help you to get started if you are not familiar with its capabilities and the basics of using it in applications. This chapter described what Enterprise Library is, where you can get it, and how it can make it much easier to manage your crosscutting concerns. This book concentrates on the application blocks in Enterprise Library that "do stuff" (as opposed to those that "wire up stuff"). The blocks we concentrate on in this book include the Data Access, Exception Handling, Transient Fault Handling, Semantic Logging, Logging, and Validation Application Blocks.

The aim of this chapter was also to help you get started with Enterprise Library by explaining how you deploy and reference the assemblies it contains, how you configure your applications to use Enterprise Library, how you instantiate Enterprise Library objects, and the example applications we provide. Some of the more advanced features and configuration options were omitted so that you may concentrate on the fundamental requirements. However, the Enterprise Library contains substantial reference documentation, samples, a reference implementation, and other resources that will guide you as you explore these more advanced features.

More Information

All links in this book are accessible from the book's online bibliography on MSDN at *http://aka.ms/el6biblio*.

If you want to know more about how you can use Unity and the Policy Injection Application Block, check out the *Dependency Injection with Unity* guide. This guide is also available on MSDN: *Developer's Guide to Dependency Injection Using Unity*.

Many scenarios supported by these blocks are now better supported by the .NET platform. Our deprecation philosophy is outlined in *this post* by the Librarian. The V6 project was focused on ensuring Enterprise Library's close alignment to the current platform (.NET framework 4.5) with a goal of reducing Enterprise Library's footprint by leveraging platform capabilities and improvements. For more details, see the *Migration Guide*.

You can download the Nucleus Research 2009 Report on Microsoft patterns & practices, which reviews the key components, benefits, and includes direct feedback from software architects and developers who have adopted patterns & practices deliverables in their projects and products from *http://www.microsoft.com/en-us/download/confirmation.aspx?id=11522* .

For more information about the topics covered in this chapter, see the *Enterprise Library Reference Documentation*.

From the perspective of Enterprise Library, the great advantage of NuGet is that it automatically adds everything that you need to use a block (including any dependencies) to your project in one, easy step. If you've not used NuGet before, you can find out more at "*Welcome to NuGet Docs*."

For more information about side-by-side operation and other deployment issues, see the *Enterprise Library Reference Documentation*.

For more information about these legacy approaches, you should read the reference documentation and the *Developer's Guide for Enterprise Library 5.0*.

Unity 3 now supports the registration by convention to make it easier to do so. See the *Dependency Injection with Unity* guide for more info.

General Links:

- There are resources to help if you're getting started with Enterprise Library, and there's help for existing users as well (such as the breaking changes and migration information for previous versions) available on the Enterprise Library Community Site at *http://www.codeplex.com/entlib/*.
- For more details about the Enterprise Library Application Blocks, see the *Enterprise Library Reference Documentation* and the *Enterprise Library 6 Class Library*.
- You can download and work through the Hands-On Labs for Enterprise Library, which are available at *http://aka.ms/el6hols*.
- You can download the Enterprise Library binaries, source code, Reference Implementation, or Quick-Starts from the Microsoft Download Center at *http://go.microsoft.com/fwlink/p/?LinkID=290898*.
- To help you understand how you can use Enterprise Library application blocks, we provide a series of simple example applications that you can run and examine. To obtain the example applications, go to *http://go.microsoft.com/fwlink/p/?LinkID=304210*.

2 Much ADO about Data Access

Using the Data Access Application Block

Introduction

When did you last write an enterprise-level application where you didn't need to handle data? And when you were handling data there was a good chance it came from some kind of relational database. Working with databases is the single most common task most enterprise applications need to accomplish, so it's no surprise that the Data Access Application Block is the most widely used of all of the Enterprise Library blocks—and no coincidence that we decided to cover it in the first of the application block chapters in this book.

A great many of the millions of Enterprise Library users around the world first cut their teeth on the Data Access block. Why? Because it makes it easy to implement the most commonly used data access operations without needing to write the same repetitive code over and over again, and without having to worry about which database the application will target. As long as there is a Data Access block provider available for your target database, you can use the same code to access the data. You don't need to worry about the syntax for parameters, the idiosyncrasies of the individual data access methods, or the different data types that are returned.

This means that it's also easy to switch your application to use a different database, without having to rewrite code, recompile, and redeploy. Administrators and operators can change the target database to a different server; and even to a different database (such as moving from Oracle to Microsoft SQL Server or the reverse), without affecting the application code. In the current release, the Data Access Application Block contains providers for SQL Server, and SQL Server Compact Edition. Support for Oracle is deprecated in this release. There are also third-party providers available for the IBM DB2, MySql, Oracle (ODP.NET), PostgreSQL, and SQLite databases. For more information on these, see the *Enterprise Library Contrib* site.

What Does the Data Access Application Block Do?

The Data Access Application Block abstracts the actual database you are using, and exposes a series of methods that make it easy to access that database to perform common tasks. It is designed to simplify the task of calling stored procedures, but also provides full support for the use of parameterized SQL statements. As an example of how easy the block is to use, when you want to fill a **DataSet** you simply create an instance of the appropriate **Database** class, use it to get an appropriate command instance (such as **DbCommand**), and pass this to the **ExecuteDataSet** method of the **Database** class. You don't need to create a **DataAdapter** or call the **Fill** method. The **ExecuteDataSet** method manages the connection, and carries out all the tasks required to populate your **DataSet**. In a similar way, the **Database** class allows you to obtain a **DataReader**, execute commands directly, and update the database from a **DataSet**. The block also supports transactions to help you manage multiple operations that can be rolled back if an error occurs.

If you want an Object/ Relational Mapping solution, you should consider using the ADO.NET Entity Framework.

In addition to the more common approaches familiar to users of ADO.NET, the Data Access block also provides techniques for asynchronous data access for databases that support this feature, and provides the ability to return data as a sequence of objects suitable for client-side querying using techniques such as Language Integrated Query (LINQ). However, the block is not intended to be an Object/Relational Mapping (O/RM) solution. It uses mappings to relate parameters and relational data with the properties of objects, but does not implement an O/RM modeling solution.

The major advantage of using the Data Access block, besides the simplicity achieved through the encapsulation of the boilerplate code that you would otherwise need to write, is that it provides a way to create provider-independent applications that can easily be moved to use a different source database type. In most cases, unless your code takes advantage of methods specific to a particular database, the only change required is to update the contents of your configuration file with the appropriate connection string. You don't have to change the way you specify queries (such as SQL statements or stored procedure names), create and populate parameters, or handle return values. This also means reduced requirements for testing.

Data Operations Supported by the Data Access Block

The following table lists by task the most commonly used methods that the Data Access Application Block exposes to retrieve and update data. Some of the method names will be familiar to those used to using ADO.NET directly.

Task	Methods
Filling a DataSet and updating the database from a DataSet.	**ExecuteDataSet**. Creates, populates, and returns a DataSet. **LoadDataSet**. Populates an existing DataSet. **UpdateDataSet**. Updates the database using an existing DataSet.
Reading multiple data rows.	**ExecuteReader**. Creates and returns a provider-independent DbDataReader instance.
Executing a Command.	**ExecuteNonQuery**. Executes the command and returns the number of rows affected. Other return values (if any) appear as output parameters. **ExecuteScalar**. Executes the command and returns a single value.
Retrieving data as a sequence of objects.	**ExecuteSprocAccessor**. Returns data selected by a stored procedure as a sequence of objects for client-side querying. **ExecuteSqlStringAccessor**. Returns data selected by a SQL statement as a sequence of objects for client-side querying.
Retrieving XML data (SQL Server only).	**ExecuteXmlReader**. Returns data as a series of XML elements exposed through an XmlReader. Note that this method is specific to the SqlDatabase class (not the underlying Database class).
Creating a Command.	**GetStoredProcCommand**. Returns a command object suitable for executing a stored procedure. **GetSqlStringCommand**. Returns a command object suitable for executing a SQL statement (which may contain parameters).
Working with Command parameters.	**AddInParameter**. Creates a new input parameter and adds it to the parameter collection of a Command. **AddOutParameter**. Creates a new output parameter and adds it to the parameter collection of a command. **AddParameter**. Creates a new parameter of the specific type and direction and adds it to the parameter collection of a command. **GetParameterValue**. Returns the value of the specified parameter as an Object type. **SetParameterValue**. Sets the value of the specified parameter.
Working with transactions.	**CreateConnection**. Creates and returns a connection for the current database that allows you to initiate and manage a transaction over the connection.

You can see from this table that the Data Access block supports almost all of the common scenarios that you will encounter when working with relational databases. Each data access method also has multiple overloads, designed to simplify usage and integrate—when necessary—with existing data transactions. In general, you should choose the overload you use based on the following guidelines:

- Overloads that accept an ADO.NET **DbCommand** object provide the most flexibility and control for each method.
- Overloads that accept a stored procedure name and a collection of values to be used as parameter values for the stored procedure are convenient when your application calls stored procedures that require parameters.
- Overloads that accept a **CommandType** value and a string that represents the command are convenient when your application executes inline SQL statements, or stored procedures that require no parameters.
- Overloads that accept a transaction allow you to execute the method within an existing transaction.
- If you use the **SqlDatabase** type, you can execute several of the common methods asynchronously by using the **Begin** and **End** versions of the methods.
- You can use the **Database** class to create **Accessor** instances that execute data access operations both synchronously and asynchronously, and return the results as a series of objects suitable for client-side querying using technologies such as LINQ.

How Do I Use the Data Access Block?

Before you start to use the Data Access block, you must add it to your application. You configure the block to specify the databases you want to work with, and add the relevant assemblies to your project. Then you can create instances of these databases in your code and use them to read and write data.

Adding the Data Access Application Block to Your Project

The first step in using the Data Access block is to add the **EnterpriseLibrary.Data** NuGet package to your project. This adds all of the required assemblies and adds the required references. If you plan to use a SQL CE database with the block, you should also add the **EnterpriseLibrary.Data.SqlCe** NuGet package. NuGet also adds references to all of the relevant assemblies in your project.

Configuring the Block and Referencing the Required Assemblies

The next step in using the Data Access block is to configure the databases you want to access. The block makes use of the standard **<connectionStrings>** section of the App.config, Web.config, or other configuration file to store the individual database connection strings, with the addition of a small Enterprise Library-specific section that defines which of the configured databases is the default. The following is an example configuration file that shows a connection string and the setting that defines the default database.

```
<?xml version="1.0"?>
<configuration>
  <configSections>
    <section name="dataConfiguration"
             type="Microsoft.Practices.EnterpriseLibrary.Data.Configuration.
                   DatabaseSettings, Microsoft.Practices.EnterpriseLibrary.Data"
             requirePermission="true"/>
  </configSections>
  <dataConfiguration defaultDatabase="ExampleDatabase"/>

  <connectionStrings>
    <add name="ExampleDatabase" connectionString="..."
        providerName="System.Data.SqlClient" />
    <add name="AsyncExampleDatabase" connectionString="..." providerName="System.Data.SqlClient" />
  </connectionStrings>
  ...
</configuration>
```

Optionally, you can use the Enterprise Library configuration tool to configure these settings.

If you are working with an Oracle database, you can use the Oracle provider included with Enterprise Library and the ADO.NET Oracle provider, which requires you to reference or add the assembly **System.Data.OracleClient.dll**. However, keep in mind that the **OracleClient** provider is deprecated in version 4.0 of the .NET Framework, and support for this provider is deprecated in Enterprise Library 6. Although support for the **OracleClient** provider is included in Enterprise Library 6, for future development you should consider choosing a different implementation of the **Database** class that uses a different Oracle driver, such as that available from the *Enterprise Library Contrib* site.

To make it easier to use the objects in the Data Access block, you can add references to the relevant namespaces, such as Microsoft.Practices.EnterpriseLibrary.Data and Microsoft.Practices.EnterpriseLibrary.Data.Sql to your project.

Creating Database Instances

You can use a variety of techniques to obtain a **Database** instance for the database you want to access. The section "Instantiating and Using Enterprise Library Objects" in Chapter 1, "Introduction" describes the different approaches you can use. The examples you can download for this chapter use the simplest approach: calling the **CreateDefault** or **Create** method of the **DatabaseProviderFactory** class, as shown here, and storing these instances in application-wide variables so that they can be accessed from anywhere in the code.

```
// Configure the DatabaseFactory to read its configuration from the .config file
DatabaseProviderFactory factory = new DatabaseProviderFactory();

// Create the default Database object from the factory.
// The actual concrete type is determined by the configuration settings.
Database defaultDB = factory.CreateDefault();

// Create a Database object from the factory using the connection string name.
Database namedDB = factory.Create("ExampleDatabase");
```

The following snippet shows the configuration settings in the app.config file in the project.

```
<?xml version="1.0"?>
<configuration>
  <configSections>
    <section name="dataConfiguration" type=".../>
  </configSections>
  <dataConfiguration defaultDatabase="ExampleDatabase"/>
  <connectionStrings>
    <add name="ExampleDatabase" connectionString="..."
         providerName="System.Data.SqlClient" />
    <add name="AsyncExampleDatabase" connectionString="..."
         providerName="System.Data.SqlClient" />
  </connectionStrings>
  ...
</configuration>
```

The code above shows how you can get an instance of the default database and a named instance (using the name in the connection strings section). Using the default database is a useful approach because you can change which of the databases defined in your configuration is the default simply by editing the configuration file, without requiring recompilation or redeployment of the application.

Notice that the code above references the database instances as instances of the **Database** base class. This is required for compatibility if you want to be able to change the database type at some later stage. However, it means that you can only use the features available across all of the possible database types (the methods and properties defined in the **Database** class).

Although you need the **dataConfiguration** section if you use the **CreateDefault** method, it doesn't need to be there if you use the **Create** method. The **Create** method reads the connection strings directly.

You may need to cast the **Database** type to a specific type such as **SqlDatabase** to access specific features. You should check that the reference returned is not null before using it.

Some features are only available in the concrete types for a specific database. For example, the **ExecuteXmlReader** method is only available in the **SqlDatabase** class. If you want to use such features, you must cast the database type you instantiate to the appropriate concrete type. The following code creates an instance of the **SqlDatabase** class.

```
// Create a SqlDatabase object from configuration using the default database.
SqlDatabase sqlServerDB = factory.CreateDefault() as SqlDatabase;
```

As alternative to using the **DatabaseProviderFactory** class, you could use the static **DatabaseFactory** façade to create your **Database** instances. You must invoke the **SetDatabaseProviderFactory** method to set the details of the default database from the configuration file.

```
DatabaseFactory.SetDatabaseProviderFactory(factory, false);
defaultDB = DatabaseFactory.CreateDatabase("ExampleDatabase");
// Uses the default database from the configuration file.
sqlServerDB = DatabaseFactory.CreateDatabase() as SqlDatabase;
```

In addition to using configuration to define the databases you will use, the Data Access block allows you to create instances of concrete types that inherit from the **Database** class directly in your code, as shown here. All you need to do is provide a connection string that specifies the appropriate ADO.NET data provider type (such as **SqlClient**).

```
// Assume the method GetConnectionString exists in your application and
// returns a valid connection string.
string myConnectionString = GetConnectionString();
SqlDatabase sqlDatabase = new SqlDatabase(myConnectionString);
```

The Example Application

Now that you have your new **Database** object ready to go, we'll show you how you can use it to perform a variety of tasks. You can download an example application (a simple console-based application) that demonstrates all of the scenarios you will see in the remainder of this chapter. You can run this directly from the bin\debug folder, or open the solution named **DataAccess** in Visual Studio to see all of the code as you run the examples.

The two connection strings for the database we provide with this example are:

```
Data Source=(localdb)\v11.0;AttachDbFilename=
|DataDirectory|\DataAccessExamples.mdf;
Integrated Security=True
Data Source=(localdb)\v11.0;Asynchronous Processing=true;
AttachDbFilename=|DataDirectory|\DataAccessExamples.mdf;
Integrated Security=True
```

If you have configured a different database using the scripts provided with the example, you may find that you get an error when you run this example. It is likely that you have an invalid connection string in your App.config file for your database.

In addition, the final example for this block uses the Distributed Transaction Coordinator (DTC) service. This service may not be set to auto-start on your machine. If you receive an error that the DTC service is not available, open the Services MMC snap-in from your Administrative Tools menu and start the service manually; then run the example again.

Reading Multiple Data Rows

One of the most common operations when working with a database is reading multiple rows of data. In a .NET application, you usually access these rows as a **DataReader** instance, or store them in a **DataTable** (usually within a **DataSet** you create). In this section we'll look at the use of the **ExecuteReader** method that returns a **DataReader**. You will see how to use a **DataSet** with the Data Access block methods later in this chapter.

The **DataSet** and **DataReader**, and related classes are still core elements of ADO.NET. However, there are newer technology options available such as Entity Framework and LINQ to SQL.

Reading Rows Using a Query with No Parameters

Simple queries consisting of an inline SQL statement or a stored procedure, which take no parameters, can be executed using the **ExecuteReader** method overload that accepts a **CommandType** value and a SQL statement or stored procedure name as a string.

The following code shows the simplest approach for a stored procedure, where you can also omit the **CommandType** parameter. The default is **CommandType.StoredProcedure** (unlike ADO.NET, where the default is **CommandType.Text**.)

```
// Call the ExecuteReader method by specifying just the stored procedure name.
using (IDataReader reader = namedDB.ExecuteReader("MyStoredProcName"))
{
  // Use the values in the rows as required.
}
To use an inline SQL statement, you must specify the appropriate CommandType
value, as shown here.
// Call the ExecuteReader method by specifying the command type
// as a SQL statement, and passing in the SQL statement.
using (IDataReader reader = namedDB.ExecuteReader(CommandType.Text,
                            "SELECT TOP 1 * FROM OrderList"))
{
  // Use the values in the rows as required - here we are just displaying them.
  DisplayRowValues(reader);
}
```

The example named *Return rows using a SQL statement with no parameters* uses this code to retrieve a **DataReader** containing the first order in the sample database, and then displays the values in this single row. It uses a simple auxiliary routine that iterates through all the rows and columns, writing the values to the console screen.

```
private static void DisplayRowValues(IDataReader reader)
{
  while (reader.Read())
  {
    for (int i = 0; i < reader.FieldCount; i++)
    {
      Console.WriteLine("{0} = {1}", reader.GetName(i), reader[i].ToString());
    }
    Console.WriteLine();
  }
}
```

The result is a list of the columns and their values in the **DataReader**, as shown here.

```
Id = 1
Status = DRAFT
CreatedOn = 01/02/2009 11:12:06
Name = Adjustable Race
LastName = Abbas
FirstName = Syed
ShipStreet = 123 Elm Street
ShipCity = Denver
ShipZipCode = 12345
ShippingOption = Two-day shipping
State = Colorado
```

Reading Rows Using an Array of Parameter Values

While you may use simple no-parameter stored procedures and SQL statements in some scenarios, it's far more common to use queries that accept input parameters that select rows or specify how the query will execute within the database server. If you use only input parameters, you pass them to the stored procedure or SQL statement directly and let the **Database** class wrap them in the appropriate **DbParameter** objects. Note that this means you must pass them in the same order as they are expected by the query, because you are not using names for these parameters—you are only supplying the actual values. The following code shows how you can execute a stored procedure that takes a single string parameter.

```
// Call the ExecuteReader method with the stored procedure
// name and an Object array containing the parameter values.
using (IDataReader reader = defaultDB.ExecuteReader("ListOrdersByState","Colorado"))
{
  // Use the values in the rows as required - here we are just displaying them.
  DisplayRowValues(reader);
}
```

The example named *Return rows using a stored procedure with parameters* uses this code to query the sample database, and generates the following output.

```
Id = 1
Status = DRAFT
CreatedOn = 01/02/2009 11:12:06
Name = Adjustable Race
LastName = Abbas
FirstName = Syed
ShipStreet = 123 Elm Street
ShipCity = Denver
ShipZipCode = 12345
ShippingOption = Two-day shipping
State = Colorado

Id = 2
Status = DRAFT
CreatedOn = 03/02/2009 01:12:06
Name = All-Purpose Bike Stand
LastName = Abel
FirstName = Catherine
ShipStreet = 321 Cedar Court
ShipCity = Denver
ShipZipCode = 12345
ShippingOption = One-day shipping
State = Colorado
```

Reading Rows Using Queries with Named Parameters

The technique in the previous example of supplying just an array of parameter values is easy and efficient, but has some limitations. It does not allow you to specify the direction (such as input or output), or the data type—which may be an issue if the data type of a parameter does not exactly match (or cannot be implicitly converted into) the correct type discovered for a stored procedure. If you create an array of parameters for your query, you can specify more details about the types of the parameters and the way they should be used.

In addition, some database systems allocate parameters used in SQL statements or stored procedures simply by position. However, many database systems, such as SQL Server, allow you to use named parameters. The database matches the names of the parameters sent with the command to the names of the parameters defined in the SQL statement or stored procedure. This means that you are not confined to adding parameters to your command in a specific order. However, be aware that if you use named parameters and then change the database type to one that does not support named parameters, any parameters that are supplied out of order will probably cause errors. (This may be difficult to detect if all of the parameters are of the same data type!)

To work with named parameters or parameters of defined types, you must access the **Command** object that will be used to execute the query, and manipulate its collection or parameters. The Data Access block makes it easy to create and access the **Command** object by using two methods of the **Database** class: **GetSqlStringCommand** and **GetStoredProcCommand**. These methods return an instance of the appropriate command class for the configured database as a provider-independent **DbCommand** type reference.

After you create the appropriate type of command, you can use the many variations of the **Database** methods to manipulate the collection of parameters. You can add parameters with a specific direction using the **AddInParameter** or **AddOutParameter** method, or by using the **AddParameter** method and providing a value for the **ParameterDirection** parameter. You can change the value of existing parameters already added to the command using the **GetParameterValue** and **SetParameterValue** methods.

The following code shows how easy it is to create a command, add an input parameter, and execute both a SQL statement and a stored procedure. Notice how the code specifies the command to which the **Database** class should add the parameter, the name, the data type, and the value of the new parameter.

```
// Read data with a SQL statement that accepts one parameter prefixed with @.
string sqlStatement = "SELECT TOP 1 * FROM OrderList WHERE State LIKE @state";

// Create a suitable command type and add the required parameter.
using (DbCommand sqlCmd = defaultDB.GetSqlStringCommand(sqlStatement))
{
  defaultDB.AddInParameter(sqlCmd, "state", DbType.String, "New York");

  // Call the ExecuteReader method with the command.
  using (IDataReader sqlReader = defaultDB.ExecuteReader(sqlCmd))
  {
    DisplayRowValues(sqlReader);
  }
}

// Now read the same data with a stored procedure that accepts one parameter.
string storedProcName = "ListOrdersByState";

// Create a suitable command type and add the required parameter.
using (DbCommand sprocCmd = defaultDB.GetStoredProcCommand(storedProcName))
{
  defaultDB.AddInParameter(sprocCmd, "state", DbType.String, "New York");

  // Call the ExecuteReader method with the command.
  using (IDataReader sprocReader = defaultDB.ExecuteReader(sprocCmd))
  {
    DisplayRowValues(sprocReader);
  }
}
```

Anytime you pass in a SQL string, you should consider carefully if you need to validate it for a possible SQL injection attack.

The example named ***Return rows using a SQL statement or stored procedure with named parameters*** uses the code you see above to execute a SQL statement and a stored procedure against the sample database. The code provides the same parameter value to each, and both queries return the same single row, as shown here.

```
Id = 4
Status = DRAFT
CreatedOn = 07/02/2009 05:12:06
Name = BB Ball Bearing
LastName = Abel
FirstName = Catherine
ShipStreet = 888 Main Street
ShipCity = New York
ShipZipCode = 54321
ShippingOption = Three-day shipping
State = New York
```

Retrieving Data as Objects

Modern programming techniques typically concentrate on data as objects. This approach is useful if you use *Data Transfer Objects* (DTOs) to pass data around your application layers, implement a data access layer using O/RM techniques, or want to take advantage of new client-side data querying techniques such as LINQ.

The Data Access block is not, in itself, an O/RM solution; but it contains features that allow you to extract data using a SQL statement or a stored procedure as the query, and have the data returned to you as a sequence of objects that implements the **IEnumerable** interface. This allows you to execute queries, or obtain lists or arrays of objects that represent the original data in the database.

About Accessors

The block provides two core classes for performing this kind of query: the **SprocAccessor** and the **SqlStringAccessor**. You can create and execute these accessors in one operation using the **ExecuteSprocAccessor** and **ExecuteSqlAccessor** methods of the **Database** class, or create a new accessor directly and then call its **Execute** method.

Accessors use two other objects to manage the parameters you want to pass into the accessor (and on to the database as it executes the query), and to map the values in the rows returned from the database to the properties of the objects it will return to the client code. Figure 1 shows the overall process.

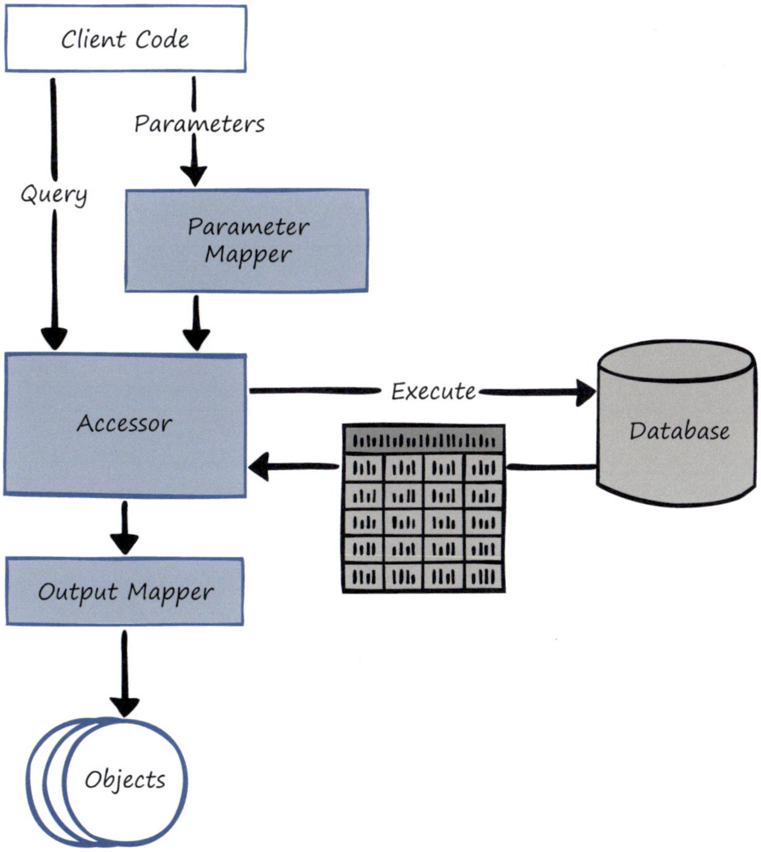

Figure 1
Overview of data accessors and the related types

The accessor will attempt to resolve the parameters automatically using a default mapper if you do not specify a parameter mapper. However, this feature is only available for stored procedures executed against SQL Server and Oracle databases. It is not available when using SQL statements, or for other databases and providers, where you must specify a custom parameter mapper that can resolve the parameters.

If you do not specify an output mapper, the block uses a default map builder class that maps the column names of the returned data to properties of the objects it creates. Alternatively, you can create a custom mapping to specify the relationship between columns in the row set and the properties of the objects.

Inferring the details required to create the correct mappings means that the default parameter and output mappers can have an effect on performance. You may prefer to create your own custom mappers and retain a reference to them for reuse when possible to maximize performance of your data access processes when using accessors.

For a full description of the techniques for using accessors, see the *Enterprise Library Reference Documentation*. This chapter covers only the simplest approach: using the **ExecuteSprocAccessor** method of the **Database** class.

Creating and Executing an Accessor

The following code shows how you can use an accessor to execute a stored procedure and then manipulate the sequence of objects that is returned. You must specify the object type that you want the data returned as—in this example it is a simple class named **Product** that has the three properties: **ID**, **Name**, and **Description**.

The stored procedure takes a single parameter that is a search string, and returns details of all products in the database that contain this string. Therefore, the code calls the **ExecuteSprocAccessor** method passing the search string as the single parameter. It specifies the **Product** class as the type of object to return, and passes to the method the name of the stored procedure to execute and the array of parameter values.

```
// Create and execute a sproc accessor that uses the default
// parameter and output mappings.
var productData = defaultDB.ExecuteSprocAccessor<Product>("GetProductList", "%bike%");

// Perform a client-side query on the returned data. Be aware that
// the orderby and filtering is happening on the client, not in the database.
var results = from productItem in productData
              where productItem.Description != null
              orderby productItem.Name
              select new { productItem.Name, productItem.Description };

// Display the results
foreach (var item in results)
{
  Console.WriteLine("Product Name: {0}", item.Name);
  Console.WriteLine("Description: {0}", item.Description);
  Console.WriteLine();
}
```

The accessor returns the data as a sequence that, in this example, the code handles using a LINQ query to remove all items where the description is empty, sort the list by name, and then create a new sequence of objects that have just the **Name** and **Description** properties. For more information on using LINQ to query sequences, see "*LINQ Query Expressions*" on MSDN.

> *Keep in mind that returning sets of data that you manipulate on the client can have an impact on performance. In general, you should attempt to return data in the format required by the client, and minimize client-side data operations.*

The example *Return data as a sequence of objects using a stored procedure* uses the code you see above to query the sample database and process the resulting rows. The output it generates is shown here.

```
Product Name: All-Purpose Bike Stand
Description: Perfect all-purpose bike stand for working on your bike at home. Quick-adjusting clamps and
steel construction.

Product Name: Bike Wash - Dissolver
Description: Washes off the toughest road grime; dissolves grease, environmentally safe. 1-liter bottle.

Product Name: Hitch Rack - 4-Bike
Description: Carries 4 bikes securely; steel construction, fits 2" receiver hitch.
```

For an example of creating an accessor and then calling the **Execute** method, see the section "Retrieving Data as Objects Asynchronously" later in this chapter.

Creating and Using Mappers

In some cases, you may need to create a custom parameter mapper to pass your parameters to the query that the accessor will execute. This typically occurs when you need to execute a SQL statement to work with a database system that does not support parameter resolution, or when a default mapping cannot be inferred due to a mismatch in the number or types of the parameters. The parameter mapper class must implement the **IParameterMapper** interface and contain a method named **AssignParameters** that takes a reference to the current **Command** instance and the array of parameters. The method simply needs to add the required parameters to the **Command** object's **Parameters** collection.

More often, you will need to create a custom output mapper. To help you do this, the block provides a class called **MapBuilder** that you can use to create the set of mappings you require between the columns of the data set returned by the query and the properties of the objects you need.

By default, the accessor will expect to generate a simple sequence of a single type of object (in our earlier example, this was a sequence of the **Product** class). However, you can use an accessor to return a more complex graph of objects if you wish. For example, you might execute a query that returns a series of **Order** objects and the related **OrderLines** objects for all of the selected orders. Simple output mapping cannot cope with this scenario, and neither can the **MapBuilder** class. In this case, you would create a result set mapper by implementing the **IResultSetMapper** interface. Your custom row set mapper must contain a method named **MapSet** that receives a reference to an object that implements the **IDataReader** interface. The method should read all of the data available through the data reader, processes it to create the sequence of objects you require, and return this sequence.

Retrieving XML Data

XML is particularly useful as standard format for exchanging data between different systems. However, XML data is typically much larger than equivalent binary data or even JSON data.

Some years ago, XML was the coolest new technology that was going to rule the world and change the way we think about data. In some ways, it did, though the emphasis on XML has receded as the relational database model continues to be the basis for most enterprise systems. However, the ability to retrieve data from a relational database as XML is useful in many scenarios, and is supported by the Data Access block.

SQL Server supports a mechanism called SQLXML that allows you to extract data as a series of XML elements, or in a range of XML document formats, by executing specially formatted SQL queries. You can use templates to precisely control the output, and have the server format the data in almost any way you require. For a description of the capabilities of SQLXML, see *SQLXML* on MSDN.

The Data Access block provides the **ExecuteXmlReader** method for querying data as XML. It takes a SQL statement that contains the **FOR XML** statement and executes it against the database, returning the result as an **XmlReader**. You can iterate through the resulting XML elements or work with them in any of the ways supported by the XML classes in the .NET Framework. However, as SQLXML is limited to SQL Server (the implementations of this type of query differ in other database systems), it is only available when you specifically use the **SqlDatabase** class (rather than the **Database** class).

The following code shows how you can obtain a **SqlDatabase** instance, specify a suitable SQLXML query, and execute it using the **ExecuteXmlReader** method.

You should check for a null reference before using the **sqlServerDB** reference.

```
// Create a SqlDatabase object from configuration using the default database.
SqlDatabase sqlServerDB = DatabaseFactory.CreateDatabase() as SqlDatabase;

// Specify a SQL query that returns XML data.
string xmlQuery = "SELECT * FROM OrderList WHERE State = @state FOR XML AUTO";

// Create a suitable command type and add the required parameter
// NB: ExecuteXmlReader is only available for SQL Server databases
using (DbCommand xmlCmd = sqlServerDB.GetSqlStringCommand(xmlQuery))
{
  xmlCmd.Parameters.Add(new SqlParameter("state", "Colorado"));
  using (XmlReader reader = sqlServerDB.ExecuteXmlReader(xmlCmd))
  {
    // Iterate through the elements in the XmlReader
    while (!reader.EOF)
    {
      if (reader.IsStartElement())
      {
        Console.WriteLine(reader.ReadOuterXml());
      }
    }
  }
}
```

The code above also shows a simple approach to extracting the XML data from the **XmlReader** returned from the **ExecuteXmlReader** method. One point to note is that, by default, the result is an XML fragment, and not a valid XML document. It is, effectively, a sequence of XML elements that represent each row in the results set. Therefore, at minimum, you must wrap the output with a single root element so that it is well-formed. For more information about using an **XmlReader**, see *"Reading XML with the XmlReader"* on MSDN.

The example *Return data as an XML fragment using a SQL Server XML query* uses the code you see above to query a SQL Server database. It returns two XML elements in the default format for a FOR XML AUTO query, with the values of each column in the data set represented as attributes, as shown here.

```
<OrderList Id="1" Status="DRAFT" CreatedOn="2009-02-01T11:12:06"
Name="Adjustable Race" LastName="Abbas" FirstName="Syed" ShipStreet="123 Elm
Street" ShipCity="Denver" ShipZipCode="12345" ShippingOption="Two-day
shipping" State="Colorado" />
<OrderList Id="2" Status="DRAFT" CreatedOn="2009-02-03T01:12:06" Name="All-
Purpose Bike Stand" LastName="Abel" FirstName="Catherine" ShipStreet="321
Cedar Court" ShipCity="Denver" ShipZipCode="12345" ShippingOption="One-day
shipping" State="Colorado" />
```

You might use this approach when you want to populate an XML document, transform the data for display, or persist it in some other form. You might use an XSLT style sheet to transform the data to the required format. For more information on XSLT, see *"XSLT Transformations"* on MSDN.

Retrieving Single Scalar Values

A common requirement when working with a database is to extract a single scalar value based on a query that selects either a single row or a single value. This is typically the case when using lookup tables or checking for the presence of a specific entity in the database. The Data Access block provides the **ExecuteScalar** method to handle this requirement. It executes the query you specify, and then returns the value of the first column of the first row of the result set as an **Object** type.

If you know that you only expect a single piece of data (one row, one column), then the **ExecuteScalar** method is the easiest way to retrieve it.

The **ExecuteScalar** method has a set of overloads similar to the **ExecuteReader** method we used earlier in this chapter. You can specify a **CommandType** (the default is **StoredProcedure**) and either a SQL statement or a stored procedure name. You can also pass in an array of **Object** instances that represent the parameters for the query. Alternatively, you can pass to the method a **Command** object that contains any parameters you require.

The following code demonstrates passing a **Command** object to the method to execute both an inline SQL statement and a stored procedure. It obtains a suitable **Command** instance from the current **Database** instance using the **GetSqlStringCommand** and **GetStoredProcCommand** methods. You can add parameters to the command before calling the **ExecuteScalar** method if required. However, to demonstrate the way the method works, the code here simply extracts the complete row set. The result is a single **Object** that you must cast to the appropriate type before displaying or consuming it in your code.

```
// Create a suitable command type for a SQL statement.
// NB: For efficiency, aim to return only a single value or a single row.
using (DbCommand sqlCmd =
                 defaultDB.GetSqlStringCommand("SELECT [Name] FROM States"))
{
    // Call the ExecuteScalar method of the command.
    Console.WriteLine("Result using a SQL statement: {0}",
                       defaultDB.ExecuteScalar(sqlCmd).ToString());
}

// Create a suitable command type for a stored procedure.
// NB: For efficiency, aim to return only a single value or a single row.
using (DbCommand sprocCmd = defaultDB.GetStoredProcCommand("GetStatesList"))
{
    // Call the ExecuteScalar method of the command.
    Console.WriteLine("Result using a stored procedure: {0}",
                       defaultDB.ExecuteScalar(sprocCmd).ToString());
}
```

You can see the code listed above running in the example *Return a single scalar value from a SQL statement or stored procedure*. The somewhat unexciting result it produces is shown here.

```
Result using a SQL statement: Alabama
Result using a stored procedure: Alabama
```

Asynchronous data access can improve the user experience because the UI does not freeze while the application fetches the data. It is also very useful in server applications where you can avoid blocking threads that could handle other requests, thus improving utilization. However, handling asynchronous calls requires more complex code: you must handle notifications when the call completes, and handle errors and timeouts asynchronously. You can now use the *Task Parallel Library* and the *async/await keywords* to greatly simplify some of your asynchronous code.

Retrieving Data Asynchronously

Having looked at all of the main ways you can extract data using the Data Access block, we'll move on to look at some more exciting scenarios (although many would perhaps fail to consider anything connected with data access exciting...). Databases are generally not renowned for being the fastest of components in an application—in fact many people will tell you that they are major bottleneck in any enterprise application. It's not that they are inefficient, it's usually just that they contain many millions of rows, and the queries you need to execute are relatively complex. Of course, it may just be that the query is badly written and causes poor performance, but that's a different story.

One way that applications can minimize the performance hit from data access is to perform it asynchronously. This means that the application code can continue to execute, and the user interface can remain interactive during the process. Asynchronous data access may not suit every situation, but it can be extremely useful.

For example, you might be able to perform multiple queries concurrently and combine the results to create the required data set. Or query multiple databases, and only use the data from the one that returned the results first (which is also a kind of failover feature). However, keep in mind that asynchronous data access has an effect on connection and data streaming performance over the wire. Don't expect a query that returns ten rows to show any improvement using an asynchronous approach—it is more likely to take longer to return the results!

The Data Access block provides asynchronous **Begin** and **End** versions of many of the standard data access methods, including **ExecuteReader**, **ExecuteScalar**, **ExecuteXmlReader**, and **ExecuteNonQuery**. It also provides asynchronous **Begin** and **End** versions of the **Execute** method for accessors that return data as a sequence of objects. You will see both of these techniques here.

Preparing for Asynchronous Data Access

Asynchronous processing in the Data Access block is only available for SQL Server databases. The **Database** class includes a property named **SupportsAsync** that you can query to see if the current **Database** instance does, in fact, support asynchronous operations. The example for this chapter contains a simple check for this.

One other point to note is that asynchronous data access usually involves the use of a callback or continuation that may run on a different thread from the calling code. On a client application some of this code usually needs to run in a UI thread, which will require proper inter-thread communication.

Other points to note about asynchronous data access are the following:

You can use the standard .NET methods and classes from the **System.Threading** namespace, such as wait handles and manual reset events, to manage asynchronous execution of the Data Access block methods. You can also cancel a pending or executing command by calling the **Cancel** method of the command you used to initiate the operation. For more information, see "*Asynchronous Operations*" on MSDN.

A recommended alternative approach is to use **Task** objects from the Task Parallel Library to wrap asynchronous operations. This is a more modern approach that simplifies some of the asynchronous code you need to write. For more information, see *Task Parallel Library* on MSDN.

- The **BeginExecuteReader** method does not accept a **CommandBehavior** parameter. By default, the method will automatically set the **CommandBehavior** property on the underlying reader to **CloseConnection** unless you specify a transaction when you call the method. If you do specify a transaction, it does not set the **CommandBehavior** property.
- Always ensure you call the appropriate **EndExecute** method when you use asynchronous data access, even if you do not actually require access to the results, or call the **Cancel** method on the connection. Failing to do so can cause memory leaks and consume additional system resources.
- Using asynchronous data access with the Multiple Active Results Set (MARS) feature of ADO.NET may produce unexpected behavior, and should generally be avoided.

The following sections illustrate two approaches to retrieving row set data asynchronously. The first uses a traditional approach, the second illustrates a more up to date approach using the Task Parallel Library

Asynchronous code is notoriously difficult to write, test, and debug for all edge cases, and you should only consider using it where it really can provide a performance benefit. For guidance on performance testing and setting performance goals see patterns & practices "*Performance Testing Guidance for Web Applications*." Using the Task Parallel Library and Task objects together with the async and await keywords makes it much easier to write asynchronous code.

Retrieving Row Set Data Asynchronously using BeginXXX and EndXXX Methods Directly

The following code shows how you can perform asynchronous data access to retrieve a row set from a SQL Server database. The code creates a **Command** instance and adds two parameters, and then calls the **BeginExecuteReader** method of the **Database** class to start the process. The code passes to this method a reference to the command to execute (with its parameters already added), a Lambda expression to execute when the data retrieval process completes, and a **null** value for the **AsyncState** parameter.

```
// Create command to execute stored procedure and add parameters.
DbCommand cmd = asyncDB.GetStoredProcCommand("ListOrdersSlowly");
asyncDB.AddInParameter(cmd, "state", DbType.String, "Colorado");
asyncDB.AddInParameter(cmd, "status", DbType.String, "DRAFT");

// Execute the query asynchronously specifying the command and the
// expression to execute when the data access process completes.
asyncDB.BeginExecuteReader(cmd,
   asyncResult =>
   {
     // Lambda expression executed when the data access completes.
     try
     {
       using (IDataReader reader = asyncDB.EndExecuteReader(asyncResult))
       {
         DisplayRowValues(reader);
       }
     }
     catch (Exception ex)
     {
       Console.WriteLine("Error after data access completed: {0}", ex.Message);
     }
   }, null);
```

The Lambda expression then calls the **EndExecuteReader** method to obtain the results of the query execution. At this point you can consume the row set in your application or, as the code above does, just display the values. Notice that the callback expression should handle any errors that may occur during the asynchronous operation.

You can also, of course, use the separate callback method approach instead of an inline lambda expression if you wish.

The **AsyncState** parameter can be used to pass any required state information into the callback. For example, when you use a callback, you would pass a reference to the current **Database** instance as the **AsyncState** parameter so that the callback code can call the **EndExecuteReader** (or other appropriate **End** method) to obtain the results. When you use a Lambda expression, the current **Database** instance is available within the expression and, therefore, you do not need to populate the **AsyncState** parameter.

In a Windows Forms or WPF application, you should not attempt to update the UI directly from within the Lambda expression: because the Lambda expression is running on a different thread, you should use the **Invoke** method on a form or the **Dispatcher** object in WPF.

The example *Execute a command that retrieves data asynchronously* uses the code shown above to fetch two rows from the database and display the contents. As well as the code above, it uses a simple routine that displays a "Waiting…" message every second as the code executes. The result is shown here.

```
Database supports asynchronous operations
Waiting... Waiting... Waiting... Waiting... Waiting...

Id = 1
Status = DRAFT
CreatedOn = 01/02/2009 11:12:06
Name = Adjustable Race
LastName = Abbas
FirstName = Syed
ShipStreet = 123 Elm Street
ShipCity = Denver
ShipZipCode = 12345
ShippingOption = Two-day shipping
State = Colorado

Id = 2
Status = DRAFT
CreatedOn = 03/02/2009 01:12:06
Name = All-Purpose Bike Stand
LastName = Abel
FirstName = Catherine
ShipStreet = 321 Cedar Court
ShipCity = Denver
ShipZipCode = 12345
ShippingOption = One-day shipping
State = Colorado
```

Of course, as we don't have a multi-million-row database handy to query, the example uses a stored procedure that contains a **WAIT** statement to simulate a long-running data access operation. It also uses **ManualResetEvent** objects to manage the threads so that you can see the results more clearly. Open the sample in Visual Studio, or view the Program.cs file, to see the way this is done.

Retrieving Row Set Data Asynchronously using a Task

In the current release of Enterprise Library, the asynchronous methods in the **Database** class do not natively support using **Task** objects. However, it is possible to use **Task** objects by using the **TaskFactory.FromAsync** method that converts a pair of **BeginXXX** and **EndXXX** methods to a **Task** implementation. The following code sample, which executes the same stored procedure as the previous example, illustrates how you can do this. Note how the example uses the **FromAsync** method to wrap the asynchronous operation in a **Task** instance and use the **async** and **await** keywords to simplify the synchronization.

```
static void ReadDataAsynchronouslyTask()
{
  if (!SupportsAsync(asyncDB)) return;
    DoReadDataAsynchronouslyTask().Wait();
}

private static async Task DoReadDataAsynchronouslyTask()
{
  try
  {
    // Create command to execute stored procedure and add parameters
    DbCommand cmd = asyncDB.GetStoredProcCommand("ListOrdersSlowly");
    asyncDB.AddInParameter(cmd, "state", DbType.String, "Colorado");
    asyncDB.AddInParameter(cmd, "status", DbType.String, "DRAFT");

    using (var timer = new Timer(_ => Console.Write("Waiting... ")))
    {
      timer.Change(0, 1000);

      using (var reader = await Task<IDataReader>.Factory.
                 FromAsync<DbCommand>(asyncDB.BeginExecuteReader,
                 asyncDB.EndExecuteReader, cmd, null))
                 {
                   timer.Change(Timeout.Infinite, Timeout.Infinite);
                   Console.WriteLine();
                   Console.WriteLine();
                   DisplayRowValues(reader);
                 }
    }
  }
  catch (Exception ex)
  {
    Console.WriteLine("Error while starting data access: {0}", ex.Message);
  }
}
```

You can run this example in the sample application to see that it behaves in the same way as the previous sample.

Retrieving Data as Objects Asynchronously

You can also execute data accessors asynchronously when you want to return your data as a sequence of objects rather than as rows and columns. The example *Execute a command that retrieves data as objects asynchronously* demonstrates this technique. You can create your accessor and associated mappers in the same way as shown in the previous section of this chapter, and then call the **BeginExecute** method of the accessor. This works in much the same way as when using the **BeginExecuteReader** method described in the previous example.

You pass to the **BeginExecute** method the lambda expression or callback to execute when the asynchronous data access process completes, along with the **AsyncState** and an array of **Object** instances that represent the parameters to apply to the stored procedure or SQL statement you are executing. The lambda expression or callback method can obtain a reference to the accessor that was executed from the **AsyncState** (casting it to an instance of the **DataAccessor** base type so that the code will work with any accessor implementation), and then call the **EndExecute** method of the accessor to obtain a reference to the sequence of objects the accessor retrieved from the database.

Updating Data

So far, we've looked at retrieving data from a database using the classes and methods of the Data Access block. Of course, while this is typically the major focus of many applications, you will often need to update data in your database. The Data Access block provides features that support data updates. You can execute update queries (such as INSERT, DELETE, and UPDATE statements) directly against a database using the **ExecuteNonQuery** method. In addition, you can use the **ExecuteDataSet**, **LoadDataSet**, and **UpdateDataSet** methods to populate a **DataSet** and push changes to the rows back into the database. We'll look at both of these approaches here.

Executing an Update Query

The Data Access block makes it easy to execute update queries against a database. By update queries, we mean inline SQL statements, or SQL statements within stored procedures, that use the **UPDATE**, **DELETE**, or **INSERT** keywords. You can execute these kinds of queries using the **ExecuteNonQuery** method of the **Database** class.

Like the **ExecuteReader** method we used earlier in this chapter, the **ExecuteNonQuery** method has a broad set of overloads. You can specify a **CommandType** (the default is **StoredProcedure**) and either a SQL statement or a stored procedure name. You can also pass in an array of **Object** instances that represent the parameters for the query. Alternatively, you can pass to the method a **Command** object that contains any parameters you require. There are also **Begin** and **End** versions that allow you to execute update queries asynchronously.

The following code from the example application for this chapter shows how you can use the **ExecuteNonQuery** method to update a row in a table in the database. It updates the **Description** column of a single row in the **Products** table, checks that the update succeeded, and then updates it again to return it to the original value (so that you can run the example again). The first step is to create the command and add the required parameters, as you've seen in earlier examples, and then call the **ExecuteNonQuery** method with the command as the single parameter. Next, the code changes the value of the command parameter named **description** to the original value in the database, and then executes the compensating update.

```
string oldDescription = "Carries 4 bikes securely; steel construction, fits 2\" receiver hitch.";
string newDescription = "Bikes tend to fall off after a few miles.";

// Create command to execute the stored procedure and add the parameters.
DbCommand cmd = defaultDB.GetStoredProcCommand("UpdateProductsTable");
defaultDB.AddInParameter(cmd, "productID", DbType.Int32, 84);
defaultDB.AddInParameter(cmd, "description", DbType.String, newDescription);

// Execute the query and check if one row was updated.
if (defaultDB.ExecuteNonQuery(cmd) == 1)
{
  // Update succeeded.
}
else
{
  Console.WriteLine("ERROR: Could not update just one row.");
}

// Change the value of the second parameter
defaultDB.SetParameterValue(cmd, "description", oldDescription);
```

```
// Execute query and check if one row was updated
if (defaultDB.ExecuteNonQuery(cmd) == 1)
{
  // Update succeeded.
}
else
{
  Console.WriteLine("ERROR: Could not update just one row.");
}
```

Notice the pattern used to execute the query and check that it succeeded. The **ExecuteNonQuery** method returns an integer value that is the number of rows updated (or, to use the more accurate term, affected) by the query. In this example, we are specifying a single row as the target for the update by selecting on the unique ID column. Therefore, we expect only one row to be updated—any other value means there was a problem.

If you are expecting to update multiple rows, you would check for a non-zero returned value. Typically, if you need to ensure integrity in the database, you could perform the update within a connection-based transaction, and roll it back if the result was not what you expected. We look at how you can use transactions with the Data Access block methods in the section "Working with Connection-Based Transactions" later in this chapter.

The example *Update data using a Command object*, which uses the code you see above, produces the following output.

```
Contents of row before update:
Id = 84
Name = Hitch Rack - 4-Bike
Description = Carries 4 bikes securely; steel construction, fits 2" receiver hitch.

Contents of row after first update:
Id = 84
Name = Hitch Rack - 4-Bike
Description = Bikes tend to fall off after a few miles.

Contents of row after second update:
Id = 84
Name = Hitch Rack - 4-Bike
Description = Carries 4 bikes securely; steel construction, fits 2" receiver hitch.
```

Working with DataSets

If you need to retrieve data and store it in a way that allows you to push changes back into the database, you will usually use a **DataSet**. The Data Access block supports simple operations on a normal (non-typed) **DataSet**, including the capability to fill a **DataSet** and then update the original database table from the **DataSet**.

To fill a **DataSet**, you use the **ExecuteDataSet** method, which returns a new instance of the **DataSet** class populated with a table containing the data for each row set returned by the query (which may be a multiple-statement batch query). The tables in this **DataSet** will have default names such as **Table**, **Table1**, and **Table2**.

If you want to load data into an existing **DataSet**, you use the **LoadDataSet** method. This allows you to specify the name(s) of the target table(s) in the **DataSet**, and lets you add additional tables to an existing **DataSet** or refresh the contents of specific tables in the **DataSet**.

Both of these methods, **ExecuteDataSet** and **LoadDataSet**, have a similar broad set of overloads to the **ExecuteReader** and other methods you've seen earlier in this chapter. You can specify a **CommandType** (the default is **StoredProcedure**) and either a SQL statement or a stored procedure name. You can also pass the values that represent the input parameters for the query. Alternatively, you can pass to the method a **Command** object that contains any parameters you require.

For example, the following code lines show how you can use the **ExecuteDataSet** method with a SQL statement; with a stored procedure and a parameter array; and with a command pre-populated with parameters. The code assumes you have created the Data Access block **Database** instance named **db**.

```
DataSet productDataSet;

// Using a SQL statement.
string sql = "SELECT CustomerName, CustomerPhone FROM Customers";
productDataSet = db.ExecuteDataSet(CommandType.Text, sql);

// Using a stored procedure and a parameter array.
productDataSet = db.ExecuteDataSet("GetProductsByCategory", "%bike%");

// Using a stored procedure and a named parameter.
DbCommand cmd = db.GetStoredProcCommand("GetProductsByCategory");
db.AddInParameter(cmd, "CategoryID", DbType.Int32, 7);
productDataSet = db.ExecuteDataSet(cmd);
```

Updating the Database from a DataSet

To update data in a database from a **DataSet**, you use the **UpdateDataSet** method, which returns a total count of the number of rows affected by the update, delete, and insert operations. The overloads of this method allow you to specify the source **DataSet** containing the updated rows, the name of the table in the database to update, and references to the three **Command** instances that the method will execute to perform UPDATE, DELETE, and INSERT operations on the specified database table.

In addition, you can specify a value for the **UpdateBehavior**, which determines how the method will apply the updates to the target table rows. You can specify one of the following values for this parameter:

- **Standard**. If the underlying ADO.NET update process encounters an error, the update stops and no subsequent updates are applied to the target table.
- **Continue**. If the underlying ADO.NET update process encounters an error, the update will continue and attempt to apply any subsequent updates.
- **Transactional**. If the underlying ADO.NET update process encounters an error, all the updates made to all rows will be rolled back.

Finally, you can—if you wish—provide a value for the **UpdateBatchSize** parameter of the **UpdateDataSet** method. This forces the method to attempt to perform updates in batches instead of sending each one to the database individually. This is more efficient, but the return value for the method will show only the number of updates made in the final batch, and not the total number for all batches. Typically, you are likely to use a batch size value between 10 and 100. You should experiment to find the most appropriate batch size; it depends on the type of database you are using, the query you are executing, and the number of parameters for the query.

You can also use the **LoadDataSet** method to load data into a typed dataset.

The examples for this chapter include one named *Fill a DataSet and update the source data,* which demonstrates the **ExecuteDataSet** and **UpdateDataSet** methods. It uses the simple overloads of the **ExecuteDataSet** and **LoadDataSet** methods to fill two **DataSet** instances, using a separate routine named **DisplayTableNames** (not shown here) to display the table names and a count of the number of rows in these tables. This shows one of the differences between these two methods. Note that the **LoadDataSet** method requires a reference to an existing **DataSet** instance, and an array containing the names of the tables to populate.

```
string selectSQL = "SELECT Id, Name, Description FROM Products WHERE Id > 90";

// Fill a DataSet from the Products table using the simple approach.
DataSet simpleDS = defaultDB.ExecuteDataSet(CommandType.Text, selectSQL);
DisplayTableNames(simpleDS, "ExecuteDataSet");

// Fill a DataSet from the Products table using the LoadDataSet method.
// This allows you to specify the name(s) for the table(s) in the DataSet.
DataSet loadedDS = new DataSet("ProductsDataSet");
defaultDB.LoadDataSet(CommandType.Text, selectSQL, loadedDS,
                      new string[] { "Products" });
DisplayTableNames(loadedDS, "LoadDataSet");
```

This produces the following result.

```
Tables in the DataSet obtained using the ExecuteDataSet method:
 - Table named 'Table' contains 6 rows.

Tables in the DataSet obtained using the LoadDataSet method:
 - Table named 'Products' contains 6 rows.
```

The example then accesses the rows in the **DataSet** to delete a row, add a new row, and change the **Description** column in another row. After this, it displays the updated contents of the **DataSet** table.

```
// get a reference to the Products table in the DataSet.
DataTable dt = loadedDS.Tables["Products"];

// Delete a row in the DataSet table.
dt.Rows[0].Delete();

// Add a new row to the DataSet table.
object[] rowData = new object[] { -1, "A New Row", "Added to the table at "
                                  + DateTime.Now.ToShortTimeString() };
dt.Rows.Add(rowData);

// Update the description of a row in the DataSet table.
rowData = dt.Rows[1].ItemArray;
rowData[2] = "A new description at " + DateTime.Now.ToShortTimeString();
dt.Rows[1].ItemArray = rowData;

// Display the contents of the DatSet.
DisplayRowValues(dt);
```

This produces the following output. To make it easier to see the changes, we've omitted the unchanged rows from the listing. Of course, the deleted row does not show in the listing, and the new row has the default ID of -1 that we specified in the code above.

```
Rows in the table named 'Products':

Id = 91
Name = HL Mountain Frame - Black, 44
Description = A new description at 14:25

...

Id = -1
Name = A New Row
Description = Added to the table at 14:25
```

The next stage is to create the commands that the **UpdateDataSet** method will use to update the target table in the database. The code declares three suitable SQL statements, and then builds the commands and adds the requisite parameters to them. Note that each parameter may be applied to multiple rows in the target table, so the actual value must be dynamically set based on the contents of the **DataSet** row whose updates are currently being applied to the target table.

This means that you must specify, in addition to the parameter name and data type, the name and the version (**Current** or **Original**) of the row in the **DataSet** to take the value from. For an INSERT command, you need the current version of the row that contains the new values. For a DELETE command, you need the original value of the ID to locate the row in the table that will be deleted. For an UPDATE command, you need the original value of the ID to locate the row in the table that will be updated, and the current version of the values with which to update the remaining columns in the target table row.

```
string addSQL    = "INSERT INTO Products (Name, Description) " + "VALUES (@name, @description)";
string updateSQL = "UPDATE Products SET Name = @name, "
                 + "Description = @description WHERE Id = @id";
string deleteSQL = "DELETE FROM Products WHERE Id = @id";

// Create the commands to update the original table in the database
DbCommand insertCommand = defaultDB.GetSqlStringCommand(addSQL);
defaultDB.AddInParameter(insertCommand, "name", DbType.String, "Name",
                         DataRowVersion.Current);
defaultDB.AddInParameter(insertCommand, "description", DbType.String,
                         "Description", DataRowVersion.Current);

DbCommand updateCommand = defaultDB.GetSqlStringCommand(updateSQL);
defaultDB.AddInParameter(updateCommand, "name", DbType.String, "Name",
                         DataRowVersion.Current);
defaultDB.AddInParameter(updateCommand, "description", DbType.String,
                         "Description", DataRowVersion.Current);
defaultDB.AddInParameter(updateCommand, "id", DbType.String, "Id",
                         DataRowVersion.Original);

DbCommand deleteCommand = defaultDB.GetSqlStringCommand(deleteSQL);
defaultDB.AddInParameter(deleteCommand, "id", DbType.Int32, "Id",
                         DataRowVersion.Original);
```

Finally, you can apply the changes by calling the **UpdateDataSet** method, as shown here.

```
// Apply the updates in the DataSet to the original table in the database.
int rowsAffected = defaultDB.UpdateDataSet(loadedDS, "Products",
                         insertCommand, updateCommand, deleteCommand,
                         UpdateBehavior.Standard);
Console.WriteLine("Updated a total of {0} rows in the database.",
                         rowsAffected);
```

The code captures and displays the number of rows affected by the updates. As expected, this is three, as shown in the final section of the output from the example.

```
Updated a total of 3 rows in the database.
```

Managing Connections

For many years, developers have fretted about the ideal way to manage connections in data access code. Connections are scarce, expensive in terms of resource usage, and can cause a big performance hit if not managed correctly. You must obviously open a connection before you can access data, and you should make sure it is closed after you have finished with it. However, if the operating system does actually create a new connection, and then closes and destroys it every time, execution in your applications would flow like molasses.

The Data Access Application Block helps you to manage your database connections automatically.

Instead, ADO.NET holds a pool of open connections that it hands out to applications that require them. Data access code must still go through the motions of calling the methods to create, open, and close connections, but ADO.NET automatically retrieves connections from the connection pool when possible, and decides when and whether to actually close the underlying connection and dispose it. The main issues arise when you have to decide when and how your code should call the Close method. The Data Access block helps to resolve these issues by automatically managing connections as far as is reasonably possible.

When you use the Data Access block to retrieve a **DataSet**, the **ExecuteDataSet** method automatically opens and closes the connection to the database. If an error occurs, it will ensure that the connection is closed. If you want to keep a connection open, perhaps to perform multiple operations over that connection, you can access the **ActiveConnection** property of your **DbCommand** object and open it before calling the **ExecuteDataSet** method. The **ExecuteDataSet** method will leave the connection open when it completes, so you must ensure that your code closes it afterwards.

In contrast, when you retrieve a **DataReader** or an **XmlReader**, the **ExecuteReader** method (or, in the case of the **XmlReader**, the **ExecuteXmlReader** method) must leave the connection open so that you can read the data. The **ExecuteReader** method sets the **CommandBehavior** property of the reader to **CloseConnection** so that the connection is closed when you dispose the reader. Commonly, you will use a **using** construct to ensure that the reader is disposed, as shown here:

```
using (IDataReader reader = db.ExecuteReader(cmd))
{
  // use the reader here
}
```

This code, and code later in this section, assumes you have created the Data Access block ***Database*** *instance named* ***db*** *and a* ***DbCommand*** *instance named* ***cmd****.*

Typically, when you use the **ExecuteXmlReader** method, you will explicitly close the connection after you dispose the reader. This is because the underlying **XmlReader** class does not expose a **CommandBehavior** property. However, you should still use the same approach as with a **DataReader** (a using statement) to ensure that the **XmlReader** is correctly closed and disposed.

```
using (XmlReader reader = db.ExecuteXmlReader(cmd))
{
  // use the reader here
}
```

Finally, if you want to be able to access the connection your code is using, perhaps to create connection-based transactions in your code, you can use the Data Access block methods to explicitly create a connection for your data access methods to use. This means that you must manage the connection yourself, usually through a **using** statement as shown below, which automatically closes and disposes the connection:

```
using (DbConnection conn = db.CreateConnection())
{
  conn.Open();
  try
  {
    // perform data access here
  }
  catch
  {
    // handle any errors here
  }
}
```

Working with Connection-Based Transactions

A common requirement in many applications is to perform multiple updates to data in such a way that they all succeed, or can all be undone (rolled back) to leave the databases in a valid state that is consistent with the original content. The traditional example is when your bank carries out a monetary transaction that requires them to subtract a payment from one account and add the same amount to another account (or perhaps slightly less, with the commission going into their own account).

Transactions should follow the four ACID principles. These are **Atomicity** (all of the tasks of a transaction are performed or none of them are), **Consistency** (the database remains in a consistent state before and after the transaction), **Isolation** (other operations cannot access or see the data in an intermediate state during a transaction), and **Durability** (the results of a successful transaction are persisted and will survive system failure).

You can execute transactions when all of the updates occur in a single database by using the features of your database system (by including the relevant commands such as BEGIN TRANSACTION and ROLLBACK TRANSACTION in your stored procedures). ADO.NET also provides features that allow you to perform connection-based transactions over a single connection. This allows you to perform multiple actions on different tables in the same database, and manage the commit or rollback in your data access code.

All of the methods of the Data Access block that retrieve or update data have overloads that accept a reference to an existing transaction as a **DbTransaction** type. As an example of their use, the following code explicitly creates a transaction over a connection. It assumes you have created the Data Access block **Database** instance named **db** and two **DbCommand** instances named **cmdA** and **cmdB**.

The **DbTransaction** class enables you to write code that requires transactions and is provider independent.

```
using (DbConnection conn = db.CreateConnection())
{
  conn.Open();
  DbTransaction trans = conn.BeginTransaction();

  try
  {
    // execute commands, passing in the current transaction to each one
    db.ExecuteNonQuery(cmdA, trans);
    db.ExecuteNonQuery(cmdB, trans);
    trans.Commit();    // commit the transaction
  }
  catch
  {
    trans.Rollback();  // rollback the transaction
  }
}
```

The examples for this chapter include one named *Use a connection-based transaction*, which demonstrates the approach shown above. It starts by displaying the values of two rows in the **Products** table, and then uses the **ExecuteNonQuery** method twice to update the **Description** column of two rows in the database within the context of a connection-based transaction. As it does so, it displays the new description for these rows. Finally, it rolls back the transaction, which restores the original values, and then displays these values to prove that it worked.

```
Contents of rows before update:

Id = 53
Name = Half-Finger Gloves, L
Description = Full padding, improved finger flex, durable palm, adjustable
closure.

Id = 84
Name = Hitch Rack - 4-Bike
Description = Carries 4 bikes securely; steel construction, fits 2" receiver
hitch.
```

```
--------------------------------------------------------------------------------
Updated row with ID = 53 to 'Third and little fingers tend to get cold.'.
Updated row with ID = 84 to 'Bikes tend to fall off after a few miles.'.
--------------------------------------------------------------------------------

Contents of row after rolling back transaction:

Id = 53
Name = Half-Finger Gloves, L
Description = Full padding, improved finger flex, durable palm, adjustable
closure.

Id = 84
Name = Hitch Rack - 4-Bike
Description = Carries 4 bikes securely; steel construction, fits 2" receiver
hitch.
```

Working with Distributed Transactions

If you need to access different databases as part of the same transaction (including databases on separate servers), of if you need to include other data sources such as Microsoft Message Queuing (MSMQ) in your transaction, you must use a distributed transaction coordinator (DTC) mechanism such as Windows Component Services. In this case, you just perform the usual data access actions, and configure your components to use the DTC. Commonly, this is done through attributes added to the classes that perform the data access.

However, ADO.NET supports the concept of automatic transactions through the **TransactionScope** class. You can specify that a series of actions require transactional support, but ADO.NET will not generate an expensive distributed transaction until you actually open more than one connection within the transaction scope. This means that you can perform multiple transacted updates to different tables in the same database over a single connection. As soon as you open a new connection, ADO.NET automatically creates a distributed transaction (using Windows Component Services), and enrolls the original connections and all new connections created within the transaction scope into that distributed transaction. You then call methods on the transaction scope to either commit all updates, or to roll back (undo) all of them.

Distributed transactions typically consume significant system resources. The **TransactionScope** class ensures that the application only uses a full distributed transaction if necessary.

Therefore, once you create the transaction scope, you use the Data Access block methods in exactly the same way as you would outside of a transaction. You do not need to pass the transaction scope to the methods as you would when using ADO.NET methods directly. For example, the methods of the Data Access Application Block automatically detect if they are being executed within the scope of a transaction. If they are, they enlist in the transaction scope and reuse the existing connection (because opening a new one might force Component Services to start a distributed transaction), and do not close the connection when they complete. The Data Access block will close and dispose the connection when the transaction is completed.

Typically, you will use the **TransactionScope** class in the following way:

```
using (TransactionScope scope = new TransactionScope(TransactionScopeOption.RequiresNew))
{
  // perform data access here
}
```

For more details about using a DTC and transaction scope, see "*Distributed Transactions (ADO.NET)*" at and "*System.Transactions Integration with SQL Server (ADO.NET)*."

The examples for this chapter contain one named *Use a TransactionScope for a distributed transaction*, which demonstrates the use of a **TransactionScope** with the Data Access block. It performs the same updates to the **Products** table in the database as you saw in the previous example of using a connection-based transaction. However, there are subtle differences in the way this example works.

In addition, as it uses the Microsoft Distributed Transaction Coordinator (DTC) service, you must ensure that this service is running before you execute the example; depending on your operating system it may not be set to start automatically. To start the service, open the Services MMC snap-in from your Administrative Tools menu, right-click on the Distributed Transaction Coordinator service, and click **Start**. To see the effects of the **TransactionScope** and the way that it promotes a transaction, open the Component Services MMC snap-in from your Administrative Tools menu and expand the **Component Services** node until you can see the **Transaction List** in the central pane of the snap-in.

When you execute the example, it creates a new **TransactionScope** and executes the **ExecuteNonQuery** method twice to update two rows in the database table. At this point, the code stops until you press a key. This gives you the opportunity to confirm that there is no distributed transaction—as you can see if you look in the transaction list in the Component Services MMC snap-in.

After you press a key, the application creates a new connection to the database (when we used a connection-based transaction in the previous example, we just updated the parameter values and executed the same commands over the same connection). This new connection, which is within the scope of the existing **TransactionScope** instance, causes the DTC to start a new distributed transaction and enroll the existing lightweight transaction into it; as shown in Figure 2.

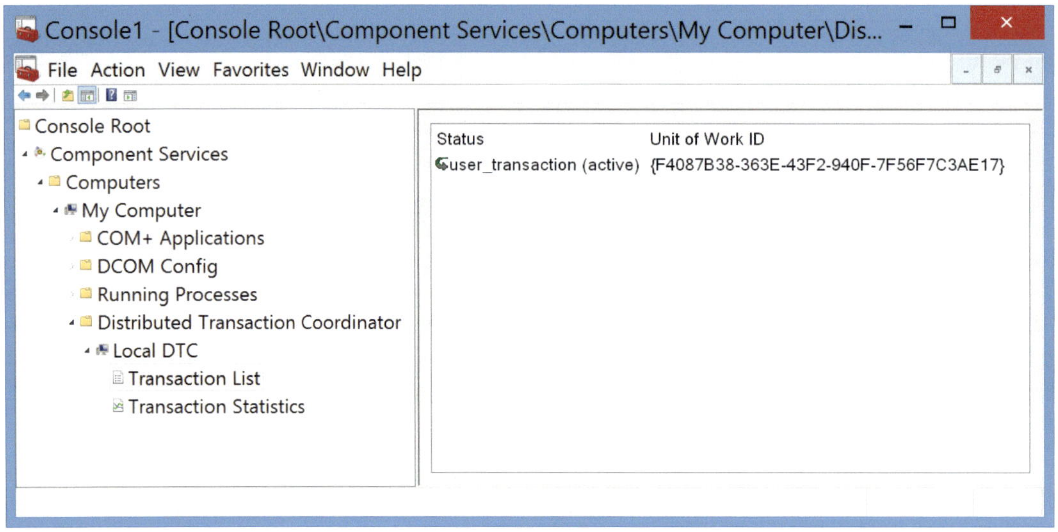

FIGURE 2
Viewing DTC transactions

The code then waits until you press a key again, at which point it exits from the using clause that created the **TransactionScope**, and the transaction is no longer in scope. As the code did not call the **Complete** method of the **TransactionScope** to preserve the changes in the database, they are rolled back automatically. To prove that this is the case, the code displays the values of the rows in the database again. This is the complete output from the example.

```
Contents of rows before update:

Id = 53
Name = Half-Finger Gloves, L
Description = Full padding, improved finger flex, durable palm, adjustable closure.

Id = 84
Name = Hitch Rack - 4-Bike
Description = Carries 4 bikes securely; steel construction, fits 2" receiver hitch.

-------------------------------------------------------------------------------
Updated row with ID = 53 to 'Third and little fingers tend to get cold.'.
No distributed transaction. Press any key to continue...

Updated row with ID = 84 to 'Bikes tend to fall off after a few miles.'.
New distributed transaction created. Press any key to continue...
-------------------------------------------------------------------------------

Contents of row after disposing TransactionScope:

Id = 53
Name = Half-Finger Gloves, L
Description = Full padding, improved finger flex, durable palm, adjustable closure.

Id = 84
Name = Hitch Rack - 4-Bike
Description = Carries 4 bikes securely; steel construction, fits 2" receiver hitch.
```

This default behavior of the **TransactionScope** ensures that an error or problem that stops the code from completing the transaction will automatically roll back changes. If your code does not seem to be updating the database, make sure you remembered to call the **Complete** method!

Extending the Block to Use Other Databases

The Data Access block contains providers for SQL Server, Oracle, and SQL Server Compact Edition. However, you can extend the block to use other databases if you wish. Writing a new provider is not a trivial task, and you may find that there is already a third party provider available for your database. For example, at the time of writing, the Enterprise Library Community Contribution site listed providers for MySql and SQLite databases. For more information, visit the *EntLib Contrib Project* site.

If you decide to create a new provider, you can create a new class derived from the Enterprise Library **Database** class and override its methods to implement the appropriate functionality. One limiting factor is that there must be an ADO.NET provider available for your database. The **Database** class in Enterprise Library relies on this to perform data access operations.

You must also be aware of the differences between database functionality, and manage these differences in your code. For example, you must handle return values, parameter prefixes (such as "@"), data type conversions, and other relevant factors. However, you can add additional methods to your provider to take advantage of features of your target database that are not available for other database types. For example, the SQL Server provider in the Data Access block exposes a method that uses the SQLXML functionality in SQL Server to extract data in XML format.

For more information on creating additional database providers for the Data Access Application Block, see the *Enterprise Library Reference Documentation*.

Summary

This chapter discussed the Data Access Application Block; one of the most commonly used blocks in Enterprise Library. The Data Access block provides two key advantages for developers and administrators. Firstly, it abstracts the database so that developers and administrators can switch the application from one type of database to another with only changes to the configuration files required. Secondly, it helps developers by making it easier to write the most commonly used sections of data access code with less effort, and it hides some of the complexity of working directly with ADO.NET.

In terms of abstracting the database, the block allows developers to write code in such a way that (for most functions) they do not need to worry which database (such as SQL Server, SQL Server CE, or Oracle) their applications will use. They write the same code for all of them, and configure the application to specify the actual database at run time. This means that administrators and operations staff can change the targeted database without requiring changes to the code, recompilation, retesting, and redeployment.

In terms of simplifying data access code, the block provides a small number of methods that encompass most data access requirements, such as retrieving a **DataSet**, a **DataReader**, a scalar (single) value, one or more values as output parameters, or a series of XML elements. It also provides methods for updating a database from a **DataSet**, and integrates with the ADO.NET **TransactionScope** class to allow a range of options for working with transactions. However, the block does not limit your options to use more advanced ADO.NET techniques, as it allows you to access the underlying objects such as the connection and the **DataAdapter**.

The chapter also described general issues such as managing connections and integration with transactions, and explored the actual capabilities of the block in more depth. Finally, we looked briefly at how you can use the block with other databases, including those supported by third-party providers.

More Information

All links in this book are accessible from the book's online bibliography on MSDN at *http://aka.ms/el6biblio*.

For more information on a variety of topics, see the *Enterprise Library Contrib site*.

For information on using *Data Transfer Objects* (DTOs) to pass data around your application layers, implement a data access layer using O/RM techniques, or want to take advantage of new client-side data querying techniques such as LINQ.

For more information on using LINQ to query sequences, see "*LINQ Query Expressions*" on MSDN.

For a description of the capabilities of SQLXML, see *SQLXML* on MSDN.

For more information about using an **XmlReader**, see "*Reading XML with the XmlReader*" on MSDN.

For more information on XSLT, see "*XSLT Transformations*" on MSDN.

You can now use the *Task Parallel Library* and the *async/await keywords* to greatly simplify some of your asynchronous code.

For more information about asynchronous data access, see "*Asynchronous Operations*" on MSDN.

For guidance on performance testing and setting performance goals see patterns & practices "*Performance Testing Guidance for Web Applications*."

For more details about using a DTC and transaction scope, see "*Distributed Transactions (ADO.NET)*" at and "*System.Transactions Integration with SQL Server (ADO.NET)*."

General Links:

- There are resources to help if you're getting started with Enterprise Library, and there's help for existing users as well (such as the breaking changes and migration information for previous versions) available on the Enterprise Library Community Site at *http://www.codeplex.com/entlib/*.
- For more details about the Enterprise Library Application Blocks, see the *Enterprise Library Reference Documentation* and the *Enterprise Library 6 Class Library*.
- You can download and work through the Hands-On Labs for Enterprise Library, which are available at *http://aka.ms/el6hols*.
- You can download the Enterprise Library binaries, source code, Reference Implementation, or Quick-Starts from the Microsoft Download Center at *http://go.microsoft.com/fwlink/p/?LinkID=290898*.
- To help you understand how you can use Enterprise Library application blocks, we provide a series of simple example applications that you can run and examine. To obtain the example applications, go to *http://go.microsoft.com/fwlink/p/?LinkID=304210*.

3 Error Management Made Exceptionally Easy

Using the Exception Handling Application Block

Introduction

Let's face it, exception handling isn't the most exciting part of writing application code. In fact, you could probably say that managing exceptions is one of those necessary tasks that absorb effort without seeming to add anything useful to your exciting new application. So why would you worry about spending time and effort actually designing a strategy for managing exceptions? Surely there are much more important things you could be doing.

In fact, a robust and well-planned exception handling plan is a vital feature of your application design and implementation. It should not be an afterthought. If you don't have a plan, you can find yourself trying to track down all kinds of strange effects and unexpected behavior in your code. And, worse than that, you may even be sacrificing security and leaving your application and systems open to attack. For example, a failure may expose error messages containing sensitive information such as: "Hi, the application just failed, but here's the name of the server and the database connection string it was using at the time." Not a great plan.

The general expectations for exception handling are to present a clear and appropriate message to users, and to provide assistance for operators, administrators, and support staff who must resolve problems that arise. For example, the following actions are usually part of a comprehensive exception handling strategy:

- Notifying the user with a friendly message
- Storing details of the exception in a production log or other repository
- Alerting the customer service team to the error
- Assisting support staff in cross-referencing the exception and tracing the cause

Applications need to deliver notifications to different groups of people such as end users, operators, and support staff. Different types and levels of information are appropriate to these different groups.

So, having decided that you probably should implement some kind of structured exception handling strategy in your code, how do you go about it? A good starting point, as usual, is to see if there are any recommendations in the form of well-known patterns that you can implement. In this case, there are. The primary pattern that helps you to build secure applications is called Exception Shielding. Exception Shielding is the process of ensuring that your application does not leak sensitive information, no matter what runtime or system event may occur to interrupt normal operation. And on a more granular level, it can prevent your assets from being revealed across layer, tier, process, or service boundaries.

Two more exception handling patterns that you should consider implementing are the Exception Logging pattern and the Exception Translation pattern. The Exception Logging pattern can help you diagnose and troubleshoot errors, audit user actions, and track malicious activity and security issues. The Exception Translation pattern describes wrapping exceptions within other exceptions specific to a layer to ensure that they actually reflect user or code actions within the layer at that time, and not some miscellaneous details that may not be useful.

In this chapter, you will see how the Enterprise Library Exception Handling block can help you to implement these patterns, and become familiar with the other techniques that make up a comprehensive exception management strategy. You'll see how to replace, wrap, and log exceptions; and how to modify exception messages to make them more useful. And, as a bonus, you'll see how you can easily implement exception shielding for Windows Communication Foundation (WCF) Web services.

When Should I Use the Exception Handling Block?

The Exception Handling block was never intended for use everywhere that you catch exceptions. The block is primarily designed to simplify exception handling and exception management at your application or layer boundaries.

The Exception Handling block allows you to configure how you want to manage exceptions, and centralize your exception handling code. It provides a selection of plug-in exception handlers and formatters that you can use, and you can even create your own custom implementations. You can use the block when you want to implement exception shielding, modify exceptions in a range of ways, or chain exceptions (for example, by logging an exception and then passing it to another layer of your application). The configurable approach means that administrators can change the behavior of the exception management mechanism simply by editing the application configuration without requiring any changes to the code, recompilation, or redeployment.

How Do I Use the Exception Handling Block?

Like all of the Enterprise Library application blocks, you start by configuring your application to use the block, as demonstrated in Chapter 1, "Introduction." You can add the block to your project by using the NuGet package manager in Visual Studio: search online in the **Manage Nuget Packages** dialog for **EnterpriseLibrary.ExceptionHandling**. Then you add one or more exception policies and, for each policy, specify the type of exception it applies to. Finally, you add one or more exception handlers to each policy. The simplest approach is a policy that specifies the base type, **Exception**, and uses one of the handlers provided with the block. However, you'll see the various handlers, and other options, demonstrated in the examples in this chapter.

There are two additional NuGet packages that you may choose to install. The **EnterpriseLibrary.ExceptionHandling.Logging** package enables you to log formatted exception information using the Logging Application Block, and the **EnterpriseLibrary.ExceptionHandling.WCF** package that includes a WCF provider. The section "Shielding Exceptions at WCF Service Boundaries" later in this chapter describes the features of this WCF provider.

What Exception Policies Do I Need?

The key to handling exceptions is to apply the appropriate policies to each type of exception. You can pretend you are playing the well-known TV quiz game that just might make you a millionaire:

Question: **How should I handle exceptions?**	
A: Wrap them	B: Replace them
C: Log and re-throw them	D: Allow them to propagate

The options in the table are the most common options, but they are not the only options. For example, you might have custom handlers, you might decide not to re-throw an exception after logging, or you might decide to simply swallow some exceptions.

You can, of course, phone a friend or ask the audience if you think it will help. However, unlike most quiz games, all of the answers are actually correct (which is why we don't offer prizes). If you answered A, B, or C, you can move on to the section "About Exception Handling Policies." However, if you answered D: Allow them to propagate, read the following section.

Allowing Exceptions to Propagate

If you cannot do anything useful when an exception occurs, such as logging exception information, modifying the exception details, or retrying the failed process, there is no point in catching the exception in the first place. Instead, you just allow it to propagate up through the call stack, and catch it elsewhere in your code—either to resolve the issue or to display error messages. Of course, at this point, you can apply an exception policy; and so you come back to how you should choose and implement an appropriate exception handling strategy.

You should consider carefully whether you should catch an exception at a particular level in your application or allow it to propagate up through the call stack. You should log all exceptions that propagate to the top of the call stack and shut the application down as gracefully as possible. For a more detailed discussion, see Eric Lippert's "Vexing exceptions" blog post and Raymond Chen's "Cleaner, more elegant, and harder to recognize" blog post.

About Exception Handling Policies

Each policy you configure for the Exception Handling block can specify one or more exception types, such as **DivideByZeroException**, **SqlException**, **Invalid-CastException**, the base class **Exception**, or any custom exception type you create that inherits from **System.Exception**. The block compares exceptions that it handles with each of these types, and chooses the one that is most specific in the class hierarchy.

For each policy, you configure:

- **One or more exception handlers** that the block will execute when a matching exception occurs. You can choose from four out-of-the-box handlers: the **Replace** handler, the **Wrap** handler, the **Logging** handler, and the **Fault Contract** exception handler. Alternatively, you can create custom exception handlers and choose these (see "Extending your Exception Handling" near the end of this chapter for more information).

- **A post-handling action** value that specifies what happens after the Exception Handling block executes the handlers you specify. Effectively, this setting tells the calling code whether to continue executing. You can choose from:
  - **NotifyRethrow** (the default). Return **true** to the calling code to indicate that it should rethrow the original exception. This allows the exception to keep its original stack trace.
  - **ThrowNewException**. The Exception Handling block will throw the exception that results from executing all of the handlers.
  - **None**. Returns **false** to the calling code to indicate that it should continue executing.

The following code sample shows how to define a policy named **MyTestExceptionPolicy** programmatically. This policy handles the three exception types—**FormatException**, **Exception**, and **InvalidCastException**—and contains a mix of handlers for each exception type. You must configure a **LogWriter** instance before you add a **Logging** exception handler to your exception handling.

```
var policies = new List<ExceptionPolicyDefinition>();

var myTestExceptionPolicy = new List<ExceptionPolicyEntry>
  {
      {
          new ExceptionPolicyEntry(typeof (FormatException),
              PostHandlingAction.NotifyRethrow,
              new IExceptionHandler[]
                  {
                    new LoggingExceptionHandler(...),
                    new ReplaceHandler(...)
                  })
      },
      {
         new ExceptionPolicyEntry(typeof (InvalidCastException), PostHandlingAction.NotifyRethrow,
              new IExceptionHandler[]
                  {
                    new LoggingExceptionHandler(...),
                  })
      },
      {
          new ExceptionPolicyEntry(typeof (Exception), PostHandlingAction.NotifyRethrow,
              new IExceptionHandler[]
                  {
                    new ReplaceHandler(...)
                  })
      }
  };

policies.Add(new ExceptionPolicyDefinition("MyTestExceptionPolicy", myTestExceptionPolicy));

ExceptionManager manager =  new ExceptionManager(policies);
```

The following code sample shows the details of a **ReplaceHandler** object. Notice how you can specify the properties for each type of exception handler. For example, you can see that the **Replace Handler** has properties for the exception message and the type of exception you want to use to replace the original exception. Also, notice that you can localize your policy by specifying the name and type of the resource containing the localized message string.

```
new ReplaceHandler(
    "Application error will be ignored and processing will continue.",
    typeof(Exception))
```

Choosing an Exception Handling Strategy

So let's get back to our quiz question, "How should I handle exceptions?" You should be able to see from the options available for exception handling policies how you can implement the common strategies for handling exceptions:

- **Replace the exception with a different one and throw the new exception**. This is an implementation of the Exception Shielding pattern. In your exception handling code, you can clean up resources or perform any other relevant processing. You use a **Replace** handler in your exception handling policy to replace the exception with a different exception containing sanitized or new information that does not reveal sensitive details about the source of the error, the application, or the operating system. Add a **Logging** handler to the exception policy if you want to log the exception. Place it before the **Replace** handler to log the original exception, or after it to log the replacement exception (if you log sensitive information, make sure your log files are properly secured). Set the post-handling action to **ThrowNewException** so that the block will throw the new exception.

- **Wrap the exception to preserve the content and then throw the new exception**. This is an implementation of the Exception Translation pattern. In your exception handling code, you can clean up resources or perform any other relevant processing. You use a **Wrap** handler in your exception-handling policy to wrap the exception within another exception and then throw the new exception so that code higher in the code stack can handle it. This approach is useful when you want to keep the original exception and its information intact, and/or provide additional information to the code that will handle the exception. Add a **Logging** handler to the exception policy if you want to log the exception. Place it before the **Wrap** handler to log the original exception, or after it to log the enclosing exception. Set the post-handling action to **ThrowNewException** so that the block will throw the new exception.

- **Log and, optionally, re-throw the original exception**. In your exception handling code, you can clean up resources or perform any other relevant processing. You use a **Logging** handler in your exception handling policy to write details of the exception to the configured logging store such as Windows Event Log or a file (an implementation of the Exception Logging pattern). If the exception does not require any further action by code elsewhere in the application (for example, if a retry action succeeds), set the post-handling action to **None**. Otherwise, set the post-handling action to **NotifyRethrow**. Your exception handling code can then decide whether to throw the exception. Alternatively, you can set it to **ThrowNewException** if you always want the Exception Handling block to throw the exception for you.

Remember that the whole idea of using the Exception Handling block is to implement a strategy made up of configurable policies that you can change without having to edit, recompile, and redeploy the application. For example, the block allows you (or an administrator) to:

- **Add, remove, and change the types of handlers** (such as the Wrap, Replace, and Logging handlers) that you use for each exception policy, and change the order in which they execute.
- **Add, remove, and change the exception types that each policy will handle**, and the types of exceptions used to wrap or replace the original exceptions.
- **Modify the target and style of logging**, including modifying the log messages, for each type of exception you decide to log. This is useful, for example, when testing and debugging applications.
- **Decide what to do after the block handles the exception**. Provided that the exception handling code you write checks the return value from the call to the Exception Handling block, the post-handling action will allow you or an administrator to specify whether the exception should be thrown. Again, this is extremely useful when testing and debugging applications.

Process or HandleException?

The Exception Handling block provides two ways for you to manage exceptions. You can use the **Process** method to execute any method in your application, and have the block automatically perform management and throwing of the exception. Alternatively, if you want to apply more granular control over the process, you can use the **HandleException** method. The following will help you to understand which approach to choose.

- The **Process** method is the most common approach, and is useful in the majority of cases. You specify either a delegate or a lambda expression that you want to execute. The Exception Handling block executes the method represented by the delegate or the body of the lambda expression, and automatically manages any exception that occurs. You will generally specify a **PostHandlingAction** of **ThrowNewException** so that the block automatically throws the exception that results from executing the exception handling policy. However, if you want the code to continue to execute (instead of throwing the exception), you can set the **PostHandlingAction** of your exception handling policy to **None**.
- The **HandleException** method is useful if you want to be able to detect the result of executing the exception handling policy. For example, if you set the **PostHandlingAction** of a policy to **NotifyRethrow**, you can use the return value of the **HandleException** method to determine whether or not to throw the exception. You can also use the **HandleException** method to pass an exception to the block and have it return the exception that results from executing the policy—which might be the original exception, a replacement exception, or the original exception wrapped inside a new exception.

You will see both the **Process** and the **HandleException** techniques described in the following examples, although most of them use the **Process** method.

Using the Process Method

The **Process** method has several overloads that make it easy to execute functions that return a value, and methods that do not. Typically, you will use the **Process** method in one of the following ways:

- To execute a routine or method that does not accept parameters and does not return a value:

```
exManager.Process(method_name, "Exception Policy Name");
```

- To execute a routine that does accept parameters but does not return a value:

```
exManager.Process(() => method_name(param1, param2), "Exception Policy Name");
```

```
To execute a routine that accepts parameters and returns a value:
```

```
var result = exManager.Process(() => method_name(param1, param2), "Exception Policy Name");
```

- To execute a routine that accepts parameters and returns a value, and to also supply a default value to be returned should an exception occur and the policy that executes does not throw the exception. If you do not specify a default value and the **PostHandlingAction** is set to **None**, the **Process** method will return **null** for reference types, zero for numeric types, or the default empty value for other types should an exception occur.

```
var result = exManager.Process(() => method-name(param1, param2), default_result_value,
                                     "Exception Policy Name");
```

- To execute code defined within the lambda expression itself:

```
exManager.Process(() =>
          {
            // Code lines here to execute application feature
            // that may raise an exception that the Exception
            // Handling block will handle using the policy named
            // in the final parameter of the Process method.
            // If required, the lambda expression defined here
            // can return a value that the Process method will
            // return to the calling code.
          },
          "Exception Policy Name");
```

The **Process** method is optimized for use with lambda expressions, which are supported in C# 3.0 on version 3.5 of the .NET Framework and in Visual Studio 2008 onwards. If you are not familiar with lambda functions or their syntax, see the "*Lambda Expressions*" article on MSDN. For a full explanation of using the **HandleException** method, see the "*Key Scenarios*" topic in the Enterprise Library Reference Documentation.

When you use the **HandleException** method, you are responsible for invoking the code that may throw an exception and for catching the exception.

Using the HandleException Method

You can use the **HandleException** method to determine whether an exception handling policy recommends that your code should rethrow an exception. There are two overloads of this method.

The section "Executing Code around Exception Handling" later in this chapter shows how you can use the **HandleException** to add additional logic around the exception management provided by the block.

The following code sample illustrates the usage of the first overload.

```
try
{
  SalaryCalculator calc = new SalaryCalculator();
  Console.WriteLine("Result is: {0}",
                   calc.RaiseArgumentOutOfRangeException("jsmith", 0));
}
catch (Exception e)
{
  if (exceptionManager.HandleException(e, "Exception Policy Name")) throw;
}
```

The following code sample illustrates the usage of the second overload. In this scenario, the **HandleException** method specifies an alternative exception to throw.

```
try
{
  SalaryCalculator calc = new SalaryCalculator();
  Console.WriteLine("Result is: {0}",
                   calc.RaiseArgumentOutOfRangeException("jsmith", 0));
}
catch (Exception e)
{
  Exception exceptionToThrow;
  if (exceptionManager.HandleException(e, "Exception Policy Name",
                                       out exceptionToThrow))
  {
    if (exceptionToThrow == null)
       throw;
    else
       throw exceptionToThrow;
  }
}
```

Diving in with a Simple Example

The code you can download for this guide contains a sample application named **ExceptionHandling** that demonstrates the techniques described in this chapter. The sample provides a number of different examples that you can run. They illustrate each stage of the process of applying exception handling described in this chapter. However, this chapter describes an iterative process of updating a single application scenario. To make it easier to see the results of each stage, we have provided separate examples for each of them.

> *If you run the examples under the Visual Studio debugger, you will find that the code halts when an exception occurs—before it is sent to the Exception Handling block. You can press F5 at this point to continue execution. Alternatively, you can run the examples by pressing Ctrl-F5 (non-debugging mode) to prevent this from happening.*

To see how you can apply exception handling strategies and configure exception handling policies, we'll start with a simple example that causes an exception when it executes. First, we need a class that contains a method that we can call from our main routine, such as the following in the **SalaryCalculator** class of the example application.

```
public Decimal GetWeeklySalary(string employeeId, int weeks)
{
  String connString = string.Empty;
  String employeeName = String.Empty;
  Decimal salary = 0;
  try
  {
    connString = ConfigurationManager.ConnectionStrings["EmployeeDatabase"].ConnectionString;

    // Access database to get salary for employee here...
    // In this example, just assume it's some large number.
    employeeName = "John Smith";
    salary = 1000000;
    return salary / weeks;
  }
  catch (Exception ex)
  {
    // provide error information for debugging
    string template = "Error calculating salary for {0}."
                    + " Salary: {1}. Weeks: {2}\n"
                    + "Data connection: {3}\n{4}";
    Exception informationException = new Exception(
                                   string.Format(template, employeeName, salary, weeks,
                                                 connString, ex.Message));
    throw informationException;
  }
}
```

The application should also raise a more specific type of exception such as **SalaryCalculationException** rather than using the **Exception** type. For more information on , see the "*Design Guidelines for Exceptions*" on MSDN.

You can see that a call to the **GetWeeklySalary** method will cause an exception of type **DivideByZeroException** when called with a value of zero for the number of weeks parameter. The exception message contains the values of the variables used in the calculation, and other information useful to administrators when debugging the application. Unfortunately, the current code has several issues. It trashes the original exception and loses the stack trace, preventing meaningful debugging. Even worse, the global exception handler for the application presents any user of the application with all of the sensitive information when an error occurs.

If you run the example for this chapter, and select option *Typical Default Behavior without Exception Shielding*, you will see this result generated by the code in the **catch** statement:

```
Exception type ExceptionHandlingExample.SalaryException was thrown.
Message: 'Error calculating salary for John Smith.
Salary: 1000000. Weeks: 0
Connection: Database=Employees;Server=CorpHQ;
User ID=admin;Password=2g$tXD76qr Attempted to divide by zero.'
Source: 'ExceptionHandlingExample'
No inner exception
```

Applying Exception Shielding

It's clear that the application as it stands has a severe security hole that allows it to reveal sensitive information. Of course, we could prevent this by not adding the sensitive information to the exception message. However, the information will be useful to administrators and developers if they need to debug the application. For example, if the data connection had failed or the database contained invalid data, they would have seen this through missing values for the employee name or salary; and they could see if the configuration file contains the correct database connection string. Alternatively, in the case shown here, they can immediately tell that the database returned the required values for the operation, but the user interface allowed the user to enter the value zero for the number of weeks.

You can create a policy that enables an administrator to control through the configuration file how much information is included in the exceptions raised by the application.

To provide this extra information, yet apply exception shielding, you may consider implementing configuration settings and custom code to allow administrators to specify when they need the additional information. However, this is exactly where the Exception Handling block comes in. You can set up an exception handling policy that administrators can modify as required, without needing to write custom code or set up custom configuration settings.

The first step is to create an exception handling policy that specifies the events you want to handle, and contains a handler that will either wrap or replace the exception containing all of the debugging information with one that contains a simple error message suitable for display to users or propagation through the layers of the application. You'll see these options implemented in the following sections. You will also see how you can log the original exception before replacing it, how you can handle specific types of exceptions, and how you can apply exception shielding to WCF services.

Wrapping an Exception

If you want to retain the original exception and the information it contains, you can wrap the exception in another exception and specify a sanitized user-friendly error message for the containing exception. This error message is available for the global error handler to display. However, code elsewhere in the application (such as code in a calling layer that needs to access and log the exception details) can access the contained exception and retrieve the information it requires before passing the exception on to another layer or to the global exception handler.

Configuring the Wrap Handler Policy

So, the first stage is to configure the exception handling policy you require. You need to add a policy that specifies the type of exception returned from your code (in this case, we'll specify the base class **Exception**), and set the **PostHandlingAction** property for this exception type to **ThrowNewException** so that the Exception Handling block will automatically throw the new exception that wraps the original exception. Then, add a **Wrap** handler to the policy, and specify the exception message and the type of exception you want to use to wrap the original exception (we chose **Exception** here again). The following code sample shows the completed configuration.

```
var exceptionShielding = new List<ExceptionPolicyEntry>
  {
      {
          new ExceptionPolicyEntry(typeof (Exception),
              PostHandlingAction.ThrowNewException,
              new IExceptionHandler[]
                {
                  new WrapHandler("Application Error. Please contact your administrator.",
                                  typeof(Exception))
                })
      }
  };
```

Initializing the Exception Handling Block

You must edit your code to use the Exception Handling block. You'll need to add references to the appropriate Enterprise Library assemblies and namespaces. The examples in this chapter demonstrate logging exception information and handling exceptions in a WCF application, as well as the basic processes of wrapping and replacing exceptions, so we'll add references to all of the assemblies and namespaces required for these tasks.

You can use NuGet to add the Exception Handling Application Block to your project. There are three NuGet packages:

- Enterprise Library - Exception Handling Application Block
- Enterprise Library - Exception Handling Application Block Logging Handler
- Enterprise Library - Exception Handling Application Block WCF Provider

If you are only wrapping and replacing exceptions in your application but not logging them, you don't need to install the Logging Handler package. If you are not using the block to shield WCF services, you don't need to install the WCF Provider package.

To make it easier to use the objects in the Exception Handling block, you can add references to the relevant namespaces to your project.

Now you can obtain an instance of the **ExceptionManager** class you'll use to perform exception management. The section "Instantiating Enterprise Library Objects" in Chapter 1, "Introduction" describes the different approaches you can use. You can choose to define the configuration information programmatically or load it from a configuration file. The examples you can download for this chapter use the programmatic approach: see the **BuildExceptionManagerConfig** method in the sample application. If you want to load the configuration data from a configuration file, you can do this by calling the **CreateManager** method of the **ExceptionPolicyFactory** class, as shown here, and storing the instance in application-wide variable so that it can be accessed from anywhere in the code.

```
// Create the default ExceptionManager object
// from the configuration settings.
ExceptionPolicyFactory policyFactory = new ExceptionPolicyFactory();
exManager = policyFactory.CreateManager();
```

Editing the Application Code to Use the New Policy

Now you can update your exception handling code to use the new policy. You have two choices. If the configured exception policy does everything you need to do, you can actually remove the **try...catch** block completely from the method that may cause an error, and—instead—use the **Process** method of the **ExceptionManager** to execute the code within the method, as shown here.

```
public Decimal GetWeeklySalary(string employeeId, int weeks)
{
  string employeeName = String.Empty;
  Decimal salary = 0;
  Decimal weeklySalary = 0;

  exManager.Process(() =>
    {
      string connString = ConfigurationManager.
                          ConnectionStrings["EmployeeDatabase"].ConnectionString;
      // Access database to get salary for employee.
      // In this example, just assume it's some large number.
      employeeName = "John Smith";
      salary = 1000000;
      weeklySalary = salary / weeks;
    },
    "ExceptionShielding");

  return weeklySalary;
}
```

The body of your logic is placed inside a lambda function and passed to the **Process** method. If an exception occurs during the execution of the expression, it is caught and handled according to the configured policy. The name of the policy to execute is specified in the second parameter of the **Process** method.

Alternatively, you can use the **Process** method in your main code to call the method of your class. This is a useful approach if you want to perform exception shielding at the boundary of other classes or objects. If you do not need to return a value from the function or routine you execute, you can create any instance you need and work with it inside the lambda expression, as shown here.

```
exManager.Process(() =>
  {
    SalaryCalculator calc = new SalaryCalculator();
    Console.WriteLine("Result is: {0}", calc.GetWeeklySalary("jsmith", 0));
  },
  "ExceptionShielding");
```

If you want to be able to return a value from the method or routine, you can use the overload of the **Process** method that returns the lambda expression value, like this.

```
SalaryCalculator calc = new SalaryCalculator();
var result = exManager.Process(() =>
             calc.GetWeeklySalary("jsmith", 0), "ExceptionShielding");
Console.WriteLine("Result is: {0}", result);
```

Notice that this approach creates the instance of the **SalaryCalculator** class outside of the **Process** method, and therefore it will not pass any exception that occurs in the constructor of that class to the exception handling policy. But when any other error occurs, the global application exception handler sees the wrapped exception instead of the original informational exception. If you run the example *Behavior After Applying Exception Shielding with a Wrap Handler*, the **catch** section now displays the following. You can see that the original exception is hidden in the Inner Exception, and the exception that wraps it contains the generic error message.

```
Exception type ExceptionHandlingExample.SalaryException was thrown.
Message: 'Application Error. Please contact your administrator.'
Source: 'Microsoft.Practices.EnterpriseLibrary.ExceptionHandling'

Inner Exception: System.Exception: Error calculating salary for John Smith.
Salary: 1000000. Weeks: 0
Connection: Database=Employees;Server=CorpHQ;User ID=admin;Password=2g$tXD76qr
Attempted to divide by zero.
   at ExceptionHandlingExample.SalaryCalculator.GetWeeklySalary(String employeeI
d, Int32 weeks) in ...\ExceptionHandling\ExceptionHandling\SalaryCalculator.cs:
line 34
   at ExceptionHandlingExample.Program.<WithWrapExceptionShielding>b__0() in
...\ExceptionHandling\ExceptionHandling\Program.cs:line 109
   at Microsoft.Practices.EnterpriseLibrary.ExceptionHandling.ExceptionManagerIm
pl.Process(Action action, String policyName)
```

This means that developers and administrators can examine the wrapped (inner) exception to get more information. However, bear in mind that the sensitive information is still available in the exception, which could lead to an information leak if the exception propagates beyond your secure perimeter. While this approach may be suitable for highly technical, specific errors, for complete security and exception shielding, you should use the technique shown in the next section to replace the exception with one that does not contain any sensitive information.

For simplicity, this example shows the principles of exception shielding at the level of the UI view. The business functionality it uses may be in the same layer, in a separate business layer, or even on a separate physical tier. Remember that you should design and implement an exception handling strategy for individual layers or tiers in order to shield exceptions on the layer or service boundaries.

Replacing an Exception

Having seen how easy it is to use exception handling policies, we'll now look at how you can implement exception shielding by replacing an exception with a different exception. To configure this scenario, simply create a policy in the same way as the previous example, but with a **Replace** handler instead of a **Wrap** handler, as shown in the following code sample.

```
var replacingException = new List<ExceptionPolicyEntry>
  {
      {
          new ExceptionPolicyEntry(typeof (Exception),
              PostHandlingAction.ThrowNewException,
              new IExceptionHandler[]
                {
                  new ReplaceHandler(
                    "Application Error. Please contact your administrator.",
                    typeof(SalaryException))
                })
      }
  };
```

When you call the method that generates an exception, you see the same generic exception message as in the previous example. However, there is no inner exception this time. If you run the example *Behavior After Applying Exception Shielding with a Replace Handler*, the Exception Handling block replaces the original exception with the new one specified in the exception handling policy. This is the result:

```
Exception type ExceptionHandlingExample.SalaryException was thrown.
Message: 'Application Error. Please contact your administrator.'
Source: 'Microsoft.Practices.EnterpriseLibrary.ExceptionHandling'
No Inner Exception
```

Logging an Exception

The previous section shows how you can perform exception shielding by replacing an exception with a new sanitized version. However, you now lose all the valuable debugging and testing information that was available in the original exception. You can preserve this information by chaining exception handlers within your exception handling policy. In other words, you add a **Logging** handler to the policy.

Logging details of the original exception enables you to capture all the valuable debugging and testing information that would otherwise be lost when you replace an exception with a new sanitized version.

That doesn't mean that the **Logging** handler is only useful as part of a chain of handlers. If you only want to log details of an exception (and then throw it or ignore it, depending on the requirements of the application), you can define a policy that contains just a **Logging** handler. However, in most cases, you will use a **Logging** handler with other handlers that wrap or replace exceptions.

The following code sample shows what happens when you add a **Logging** handler to your exception handling policy. You must create a **LogWriter** instance to use with the handler. You also need to set a few properties of the **Logging** exception handler in the **Exception Handling Settings** section:

- Specify the ID for the log event your code will generate as the **Event ID** property.
- Specify the **TextExceptionFormatter** as the type of formatter the Exception Handling block will use. Click the ellipsis (...) button in the **Formatter Type** property and select **TextExceptionFormatter** in the type selector dialog that appears.
- Set the category for the log event. The Logging block contains a default category named **General**, and this is the default for the Logging exception handler. However, if you configure other categories for the Logging block, you can select one of these from the drop-down list that is available when you click on the **Logging Category** property of the **Logging** handler.

```
var loggingAndReplacing = new List<ExceptionPolicyEntry>
  {
      {
          new ExceptionPolicyEntry(typeof (Exception), PostHandlingAction.ThrowNewException,
              new IExceptionHandler[]
                {
                  new LoggingExceptionHandler(
                      "General", 9000, TraceEventType.Error,
                      "Salary Calculations Service", 5,
                      typeof(TextExceptionFormatter), logWriter),
                  new ReplaceHandler(
                      "An application error occurred and has been logged.
                      Please contact your administrator.",
                      typeof(SalaryException))
                })
      }
  };
```

The following code sample shows the definition of the **LogWriter** instance.

```
// Formatters
TextFormatter formatter = new TextFormatter("Timestamp: ... ");

// Listeners
var flatFileTraceListener = new FlatFileTraceListener(..., formatter);
var eventLog = new EventLog("Application", ".", "Enterprise Library Logging");
var eventLogTraceListener = new FormattedEventLogTraceListener(eventLog);

// Build Configuration
var config = new LoggingConfiguration();
config.AddLogSource("General", SourceLevels.All, true)
     .AddTraceListener(eventLogTraceListener);
config.LogSources["General"].AddTraceListener(flatFileTraceListener);

// Special Sources Configuration
config.SpecialSources.LoggingErrorsAndWarnings.AddTraceListener(eventLogTraceListener);
LogWriter logWriter = new LogWriter(config);
```

In addition, if you did not already do so, you must add a reference to the Logging Application block assembly to your project and (optionally) add a **using** statement to your class, as shown here.

```
using Microsoft.Practices.EnterpriseLibrary.Logging;
```

Now, when the application causes an exception, the global exception handler continues to display the same sanitized error message. However, the **Logging** handler captures details of the original exception before the Exception Handling block policy replaces it, and writes the details to whichever logging sink you specify in the configuration for the Logging block. If you run the example *Logging an Exception to Preserve the Information it Contains*, you will see an exception like the one in Figure 1.

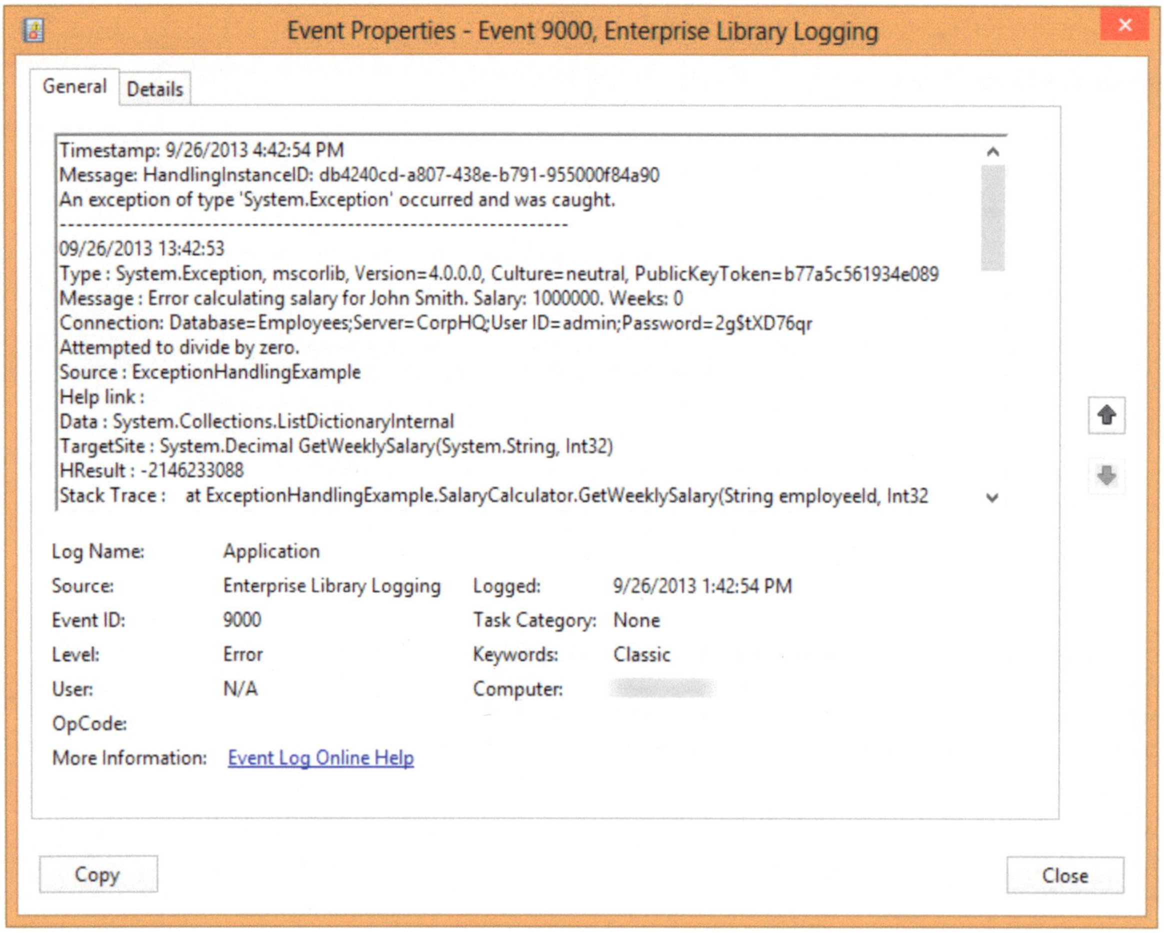

FIGURE 1
Details of the logged exception

This example shows the Exception Handling block using the default settings for the Logging block. However, as you can see in Chapter 5, "As Easy As Falling Off a Log," the Logging block is extremely configurable. So you can arrange for the Logging handler in your exception handling policy to write the information to any Windows Event Log, an email message, a database, a message queue, a text file, or a custom location using classes you create that take advantage of the application block extension points.

Shielding Exceptions at WCF Service Boundaries

You can use the Exception Handling block to implement exception handling policies for WCF services. A common scenario is to implement the Exception Shielding pattern at a WCF service boundary. The Exception Handling block contains a handler specifically designed for this (the **Fault Contract** exception handler), which maps the values in the exception to a new instance of a fault contract that you specify.

Creating a Fault Contract

A fault contract for a WCF service will generally contain just the most useful properties for an exception, and exclude sensitive information such as the stack trace and anything else that may provide attackers with useful information about the internal workings of the service. The following code shows a simple example of a fault contract suitable for use with the **Fault Contract** exception handler:

```
[DataContract]
public class SalaryCalculationFault
{
  [DataMember]
  public Guid FaultID { get; set; }

  [DataMember]
  public string FaultMessage { get; set; }
}
```

Configuring the Exception Handling Policy

The following code sample shows a sample configuration for the **Fault Contract** exception handler. This specifies the type **SalaryCalculationFault** as the target fault contract type, and the exception message that the policy will generate to send to the client. Note that, when using the **Fault Contract** exception handler, you should always set the **PostHandlingAction** property to **ThrowNewException** so that the Exception Handling block throws an exception that forces WCF to return the fault contract to the client.

```
var mappings = new NameValueCollection();
mappings.Add("FaultID", "{Guid}");
mappings.Add("FaultMessage", "{Message}");
var salaryServicePolicy = new List<ExceptionPolicyEntry>
    {
        new ExceptionPolicyEntry(typeof (Exception), PostHandlingAction.ThrowNewException,
            new IExceptionHandler[]
                {
                  new FaultContractExceptionHandler(typeof(SalaryCalculationFault), mappings)
                })
    };
```

Notice that we specified **Property Mappings** for the handler that map the **Message** property of the exception generated within the service to the **FaultMessage** property of the **SalaryCalculationFault** class, and map the unique Handling Instance ID of the exception (specified by setting the **Source** to "{**Guid**}") to the **FaultID** property.

Editing the Service Code to Use the New Policy

After you specify your fault contract and configure the **Fault Contract** exception handler, you must edit your service code to use the new exception policy. If you did not already do so, you must also add a reference to the assembly that contains the **Fault Contract** exception handler to your project and (optionally) add a **using** statement to your service class, as shown here:

```
using Microsoft.Practices.EnterpriseLibrary.ExceptionHandling.WCF;
```

Now, you can add attributes to the methods in your service class to specify the policy they should use when an exception occurs, as shown in this code:

```
[ServiceContract]
public interface IsalaryService
{
  [OperationContract]
  [FaultContract(typeof(SalaryCalculationFault))]
  decimal GetWeeklySalary(string employeeId, int weeks);
}

[DataContract]
public class SalaryCalculationFault
{
  [DataMember]
  public Guid FaultID { get; set; }

  [DataMember]
  public string FaultMessage { get; set; }
}

[ExceptionShielding("SalaryServicePolicy")]
public class SalaryService : IsalaryService
{
  public decimal GetWeeklySalary(string employeeId, int weeks)
  {
    SalaryCalculator calc = new SalaryCalculator();
    return calc.GetWeeklySalary(employeeId, weeks);
  }
}
```

The **ExceptionShielding** attribute specifies the policy to use when the service is invoked using the specified contracts. Specifying the policy programmatically may be safer than using declarative configuration because a change to the declarative configuration might affect how the contract is fulfilled.

You add the **ExceptionShielding** attribute to a service implementation class or to a service contract interface, and use it to specify the name of the exception policy to use. If you do not specify the name of a policy in the constructor parameter, or if the specified policy is not defined in the configuration, the Exception Handling block will automatically look for a policy named **WCF Exception Shielding**.

You also need to invoke the static **SetExceptionManager** method on the **ExceptionPolicy** class to ensure that the policy is loaded. The following code sample (specific to ASP.NET hosting) shows a method invoked from the **Application_Start** method that does this:

```
private void ConfigureExceptionHandlingBlock()
{
  var policies = new List<ExceptionPolicyDefinition>();

  var mappings = new NameValueCollection();
  mappings.Add("FaultID", "{Guid}");
  mappings.Add("FaultMessage", "{Message}");

  var salaryServicePolicy = new List<ExceptionPolicyEntry>
    {
        new ExceptionPolicyEntry(typeof (Exception),
            PostHandlingAction.ThrowNewException,
            new IExceptionHandler[]
              {
                new FaultContractExceptionHandler(typeof(SalaryCalculationFault), mappings)
              })
    };
  policies.Add(new ExceptionPolicyDefinition("SalaryServicePolicy", salaryServicePolicy));

  ExceptionManager exManager = new ExceptionManager(policies);
  ExceptionPolicy.SetExceptionManager(exManager);
}
```

The Fault Contract Exception Handler

The Exception Handling block executes the **Fault Contract** exception handler that you specify in your policy when an exception occurs. Effectively, the **Fault Contract** handler is a specialized version of the **Replace** handler. It takes the original exception, generates an instance of the fault contract, populates it with values from the exception, and then throws a **FaultException<***YourFaultContractType***>** exception. The handler performs the following actions:

- It generates a new instance of the fault contract class you specify for the **FaultContractType** property of the **Fault Contract** exception handler.
- It extracts the values from the properties of the exception that you pass to the method.
- It sets the values of the new fault contract instance to the values extracted from the original exception. It uses mappings between the exception property names and the names of the properties exposed by the fault contract to assign the exception values to the appropriate properties. If you do not specify a mapping, it matches the source and target properties that have the same name.

The result is that, instead of a general service failure message, the client receives a fault message containing the appropriate information about the exception.

The example *Applying Exception Shielding at WCF Application Boundaries* uses the service described above and the Exception Handling block WCF Fault Contract handler to demonstrate exception shielding. You can run this example in one of three ways:

- Inside Visual Studio by starting it with *F5* (debugging mode) and then pressing *F5* again when the debugger halts at the exception in the **SalaryCalculator** class.
- Inside Visual Studio by right-clicking SalaryService.svc in Solution Explorer and selecting **View in Browser** to start the service, then pressing *Ctrl-F5* (non-debugging mode) to run the application.
- By starting the **SalaryService** in Visual Studio (as described in the previous bullet) and then running the executable file **ExceptionHandlingExample.exe** in the **bin\debug** folder directly.

The result is shown below. You can see that the exception raised by the **SalaryCalculator** class causes the service to return an instance of the **SalaryCalculationFault** type that contains the fault ID and fault message. However, the Exception Handling block captures this exception and replaces the sensitive information in the message with text that suggests the user contact their administrator. Research shows that users really appreciate this type of informative error message.

```
Getting salary for 'jsmith' from WCF Salary Service...
Exception type System.ServiceModel.FaultException`1[ExceptionHandlingExample.Sal
aryService.SalaryCalculationFault] was thrown.
Message: 'Service error. Please contact your administrator.'
Source: 'mscorlib'
No Inner Exception

Fault contract detail:
Fault ID: bafb7ec2-ed05-4036-b4d5-56d6af9046a5
Message: Error calculating salary for John Smith. Salary: 1000000. Weeks: 0
Connection: Database=Employees;Server=CorpHQ;User ID=admin;Password=2g$tXD76qr
Attempted to divide by zero.
```

You can also see, below the details of the exception, the contents of the original fault contract, which are obtained by casting the exception to the type **FaultException<SalaryCalculationFault>** and querying the properties. You can see that this contains the original exception message generated within the service. Look at the code in the example file, and run it, to see more details.

> *It is not recommended to include the original message in the **FaultException** as this may expose the sensitive information you are trying to hide. This example uses the original message simply to illustrate how you can populate the **FaultException**.*

Handling Specific Exception Types

So far, all of the examples have used an exception policy that handles a single exception type (in the examples, this is **Exception** so that the policy will apply to any type of exception passed to it). However, you can specify multiple exception types within a policy, and specify different handlers—including chains of handlers—for each exception type. The following code sample shows the configuration of a policy with two exception types defined for the policy named **LogAndWrap** (which is used in the next example). Each exception type has different exception handlers specified.

```
var logAndWrap = new List<ExceptionPolicyEntry>
  {
      {
          new ExceptionPolicyEntry(typeof (DivideByZeroException),
              PostHandlingAction.None,
              new IExceptionHandler[]
                {
                  new LoggingExceptionHandler("General", 9002,
                      TraceEventType.Error,
                      "Salary Calculations Service", 5,
                      typeof(TextExceptionFormatter), logWriter),
                  new ReplaceHandler("Application error will be ignored
                      and processing will continue.",
                      typeof(Exception))
                })
      },
      {
          new ExceptionPolicyEntry(typeof (Exception),
              PostHandlingAction.ThrowNewException,
              new IExceptionHandler[]
                {
                  new WrapHandler("An application error has occurred.",
                      typeof(Exception))
                })
      }
  };
```

The advantage of this capability should be obvious. You can create policies that will handle different types of exceptions in different ways and, for each exception type, can have different messages and post-handling actions as well as different handler combinations. They can add new exception types, modify the types specified, change the properties for each exception type and the associated handlers, and generally fine-tune the strategy to suit day-to-day operational requirements.

Of course, this will only work if your application code throws the appropriate exception types. If you generate informational exceptions that are all of the base type **Exception**, as we did in earlier examples in this chapter, only the handlers for that exception type will execute.

Of course, you should throw exceptions of specific types in in your application and not throw exceptions of the base type **Exception**.

Executing Code around Exception Handling

So far, all of the examples have used the Process method to execute the code that may cause an exception. They simply used the **Process** method to execute the target class method, as shown here.

```
SalaryCalculator calc = new SalaryCalculator();
var result = exManager.Process(() =>
             calc.GetWeeklySalary("jsmith", 0), "ExceptionShielding");
Console.WriteLine("Result is: {0}", result);
```

However, as you saw earlier in this chapter, the **Process** method does not allow you to detect the return value from the exception handling policy executed by the Exception Handling block (it returns the value of the method or function it executes). In some cases, though perhaps rarely, you may want to detect the return value from the exception handling policy and perform some processing based on this value, and perhaps even capture the exception returned by the Exception Handling block to manipulate it or decide whether or not to throw it in your code.

In this case, you can use the **HandleException** method to pass an exception to the block as an **out** parameter to be populated by the policy, and retrieve the Boolean result that indicates if the policy determined that the exception should be thrown or ignored.

The example *Executing Custom Code Before and After Handling an Exception*, demonstrates this approach. The **SalaryCalculator** class contains two methods in addition to the **GetWeeklySalary** method we've used so far in this chapter. These two methods, named **RaiseDivideByZeroException** and **RaiseArgumentOutOfRange-Exception**, will cause an exception of the type indicated by the method name when called.

The sample first attempts to execute the **RaiseDivideByZeroException** method, like this.

```
SalaryCalculator calc = new SalaryCalculator();
Console.WriteLine("Result is: {0}", calc.RaiseDivideByZeroException("jsmith", 0));
```

This exception is caught in the main routine using the exception handling code shown below. This creates a new **Exception** instance and passes it to the Exception Handling block as the **out** parameter, specifying that the block should use the **NotifyingRethrow** policy. This policy specifies that the block should log **DivideByZero** exceptions, and replace the message with a sanitized one. However, it also has the **PostHandlingAction** set to **None**, which means that the **HandleException** method will return false. The sample code simply displays a message and continues.

```
...
catch (Exception ex)
{
  Exception newException;
  bool rethrow = exManager.HandleException(ex, "NotifyingRethrow", out newException);
  if (rethrow)
  {
    // Exception policy could specify to throw the existing exception
    // or the new exception.
    // Code here to perform any clean up tasks required.
    if (newException == null)
       throw;
    else
       throw newException;
  }
  else
  {
    // Exception policy setting is "None" so exception is not thrown.
    // Code here to perform any other processing required.
    // In this example, just ignore the exception and do nothing.
    Console.WriteLine("Detected and ignored Divide By Zero Error " +
                      "- no value returned.");
  }
}
```

Therefore, when you execute this sample, the following message is displayed.

```
Getting salary for 'jsmith' ... this will raise a DivideByZero exception.
Detected and ignored Divide By Zero Error - no value returned.
```

The sample then continues by executing the **RaiseArgumentOutOfRangeException** method of the **Salary-Calculator** class, like this.

```
SalaryCalculator calc = new SalaryCalculator();
Console.WriteLine("Result is: {0}",
                  calc.RaiseArgumentOutOfRangeException("jsmith", 0));
```

This section of the sample also contains a **catch** section, which is—other than the message displayed to the screen—identical to that shown earlier. However, the **NotifyingRethrow** policy specifies that exceptions of type **Exception** (or any exceptions that are not of type **DivideByZeroException**) should simply be wrapped in a new exception that has a sanitized error message. The **PostHandlingAction** for the Exception type is set to **ThrowNewException**, which means that the **HandleException** method will return true. Therefore the code in the catch block will throw the exception returned from the block, resulting in the output shown here.

```
Getting salary for 'jsmith' ... this will raise an ArgumentOutOfRange exception.

Exception type System.Exception was thrown.
Message: 'An application error has occurred.'
Source: 'ExceptionHandlingExample'

Inner Exception: System.ArgumentOutOfRangeException: startIndex cannot be larger than length of string.
Parameter name: startIndex
   at System.String.InternalSubStringWithChecks(Int32 startIndex, Int32 length, Boolean fAlwaysCopy)
   at System.String.Substring(Int32 startIndex, Int32 length)
   at ExceptionHandlingExample.SalaryCalculator.RaiseArgumentOutOfRangeException
(String employeeId, Int32 weeks) in ...\ExceptionHandling\ExceptionHandling\SalaryCalculator.cs:line 57
   at ExceptionHandlingExample.Program.ExecutingCodeAroundException(Int32 positionInTitleArray) in
...\ExceptionHandling\ExceptionHandling\Program.cs:line 222
```

You can also use the static **HandleException** method of the **ExceptionPolicy** class to determine what action you should take when you handle an exception. Before you use this static method you must set the exception manager as shown in the following code sample:

```
ExceptionManager exManager;
...
ExceptionPolicy.SetExceptionManager(exManager);
```

The following code from the example *Using the static ExceptionPolicy class* demonstrates this approach.

```
try
{
  result = calc.GetWeeklySalary("jsmith", 0);
}
catch (Exception ex)
{
  Exception exceptionToThrow;
  if (ExceptionPolicy.HandleException(ex, "ExceptionShielding",
      out exceptionToThrow))
  {
    if (exceptionToThrow == null)
        throw;
    else
        throw exceptionToThrow;
  }
}
```

Assisting Administrators

You can't rely on users to report accurately all of the details of an exception. The **HandlingInstanceID** value enables you to give the end user a single piece of information that you can use to identify the specific circumstances of an exception in the application. In addition, you should consider using end-to-end tracing to correlate all of the information about a single user action.

Some would say that the Exception Handling block already does plenty to make an administrator's life easy. However, it also contains features that allow you to exert extra control over the way that exception information is made available, and the way that it can be used by administrators and operations staff. If you have ever worked in a technical support role, you'll recognize the scenario. A user calls to tell you that an error has occurred in the application. If you are lucky, the user will be able to tell you exactly what they were doing at the time, and the exact text of the error message. More likely, he or she will tell you that they weren't really doing anything, and that the message said something about contacting the administrator.

To resolve this regularly occurring problem, you can make use of the **HandlingInstanceID** value generated by the block to associate logged exception details with specific exceptions, and with related exceptions. The Exception Handling block creates a unique GUID value for the **HandlingInstanceID** of every execution of a policy. The value is available to all of the handlers in the policy while that policy is executing. The **Logging** handler automatically writes the **HandlingInstanceID** value into every log message it creates. The **Wrap** and **Replace** handlers can access the **HandlingInstanceID** value and include it in a message using the special token {**handlingInstanceID**}.

The following code sample shows how you can configure a **Logging** handler and a **Replace** handler in a policy, and include the **{handlingInstanceID}** token in the **Exception Message** property of the **Replace** handler.

```
var assistingAdministrators = new List<ExceptionPolicyEntry>
  {
      {
          new ExceptionPolicyEntry(typeof (Exception),
              PostHandlingAction.ThrowNewException,
              new IExceptionHandler[]
              {
                  new LoggingExceptionHandler("General", 9001,
                      TraceEventType.Error,
                      "Salary Calculations Service", 5,
                      typeof(TextExceptionFormatter), logWriter),
                  new ReplaceHandler("Application error.
                      Please advise your administrator and provide
                      them with this error code: {handlingInstanceID}",
                      typeof(SalaryException))
              })
      }
  };
```

Now your application can display the unique exception identifier to the user, and they can pass it to the administrator who can use it to identify the matching logged exception information. This logged information will include the information from the original exception, before the **Replace** handler replaced it with the sanitized exception. If you select the option *Providing Assistance to Administrators for Locating Exception Details* in the example application, you can see this in operation. The example displays the following details of the exception returned from the exception handling policy:

```
Exception type ExceptionHandlingExample.SalaryException was thrown.
Message: 'Application error. Please advise your administrator and provide them
with this error code: 22f759d3-8f58-43dc-9adc-93b953a4f733'
Source: 'Microsoft.Practices.EnterpriseLibrary.ExceptionHandling'
No Inner Exception
```

In a production application, you will probably show this message in a dialog of some type. One issue, however, is that users may not copy the GUID correctly from a standard error dialog (such as a message box). If you decide to use the **HandlingInstanceID** value to assist administrators, consider using a form containing a read-only text box or an error page in a web application to display the GUID value in a way that allows users to copy it to the clipboard and paste into a document or email message.

Extending Your Exception Handling

Like all of the Enterprise Library application blocks, the Exception Handling block is extensible. You can create new exception handlers and exception formatters if you need to perform specific tasks that the block does not implement by default. For example, you could create an exception handler that displays the exception details in a dialog, creates an XML message, or generates a web page. All that is required is that you implement the **IExceptionHandler** interface defined within the block. This interface contains one method named **HandleException** that the Exception Handling block will execute when you add your handler to a policy chain.

For information about adding support for declarative configuration in custom handlers, see the topic *Extending and Modifying the Exception Handling Application Block* in the Enterprise Library Reference Documentation.

The Exception Handling block uses formatters to create the message sent to the Logging block when you use a **Logging** handler. The two formatters provided are the **TextExceptionFormatter** and the **XmlExceptionFormatter**. The **TextExceptionFormatter** generates a simple text format for the error message, as you have seen in the previous sections of this chapter. The **XmlExceptionFormatter** generates, as you would expect, an XML-formatted document as the error message. You can create custom formatters if you want to control the exact format of this information. You simply create a new class that derives from the **ExceptionFormatter** base class in the Exception Handling block, and override the several methods it contains for formatting the exception information as required.

Summary

In this chapter you have seen why, when, and how you can use the Enterprise Library Exception Handling block to create and implement exception handling strategies. Poor error handling can make your application difficult to manage and maintain, hard to debug and test, and may allow it to expose sensitive information that would be useful to attackers and malicious users.

A good practice for exception management is to implement strategies that provide a controlled and decoupled approach to exception handling through configurable policies. The Exception Handling block makes it easy to implement such strategies for your applications, irrespective of their type and complexity. You can use the Exception Handling block in Web and Windows Forms applications, Web services, console-based applications and utilities, and even in administration scripts and applications hosted in environments such as SharePoint, Microsoft Office applications, or other enterprise systems.

This chapter demonstrated how you can implement common exception handling patterns, such as Exception Shielding, using techniques such as wrapping, replacing, and logging exceptions. It also demonstrated how you can handle different types of exceptions, assist administrators by using unique exception identifiers, and extend the Exception Handling block to perform tasks that are specific to your own requirements.

More Information

> All links in this book are accessible from the book's online bibliography on MSDN at *http://aka.ms/el6biblio*.

For a more detailed discussion of whether you should catch an exception at a particular level in your application or allow it to propagate up through the call stack, see Eric Lippert's "*Vexing exceptions*" blog post and Raymond Chen's "*Cleaner, more elegant, and harder to recognize*" blog post.

If you are not familiar with lambda functions or their syntax, see the "*Lambda Expressions*" article on MSDN.

For a full explanation of using the **HandleException** method, see the "*Key Scenarios*" topic in the Enterprise Library Reference Documentation.

For more information on raising specific types of exceptions, see the "*Design Guidelines for Exceptions*" on MSDN.

For information about adding support for declarative configuration in custom handlers, see the topic *Extending and Modifying the Exception Handling Application Block* in the Enterprise Library Reference Documentation.

General Links:

- There are resources to help if you're getting started with Enterprise Library, and there's help for existing users as well (such as the breaking changes and migration information for previous versions) available on the Enterprise Library Community Site at *http://www.codeplex.com/entlib/*.
- For more details about the Enterprise Library Application Blocks, see the *Enterprise Library Reference Documentation* and the *Enterprise Library 6 Class Library*.
- You can download and work through the Hands-On Labs for Enterprise Library, which are available at *http://aka.ms/el6hols*.
- You can download the Enterprise Library binaries, source code, Reference Implementation, or Quick-Starts from the Microsoft Download Center at *http://go.microsoft.com/fwlink/p/?LinkID=290898*.
- To help you understand how you can use Enterprise Library application blocks, we provide a series of simple example applications that you can run and examine. To obtain the example applications, go to *http://go.microsoft.com/fwlink/p/?LinkID=304210*.

4 Perseverance, Secret of All Triumphs

Using the Transient Fault Handling Application Block

What Are Transient Faults?

When an application uses a service, errors can occur because of temporary conditions such as intermittent service, infrastructure-level faults, network issues, or explicit throttling by the service; these types of error occur more frequently with cloud-based services, but can also occur in on-premises solutions. If you retry the operation a short time later (maybe only a few milliseconds later) the operation may succeed. These types of error conditions are referred to as transient faults. Transient faults typically occur very infrequently, and in most cases, only a few retries are necessary for the operation to succeed.

Unfortunately, there is no easy way to distinguish transient from non-transient faults; both would most likely result in exceptions being raised in your application. If you retry the operation that causes a non-transient fault (for example a "file not found" error), you most likely get the same exception raised again.

> There is no intrinsic way to distinguish between transient and non-transient faults unless the developer of the service explicitly isolated transient faults into a specified subset of exception types or error codes.

For example, with SQL Database, one of the important considerations is how you should handle client connections. This is because SQL Database can use throttling when a client attempts to establish connections to a database or run queries against it. SQL Database throttles the number of database connections for a variety of reasons, such as excessive resource usage, long-running transactions, and possible failover and load balancing actions. This can lead to the termination of existing client sessions or the temporary inability to establish new connections while the transient conditions persist. SQL Database can also drop database connections for a variety of reasons related to network connectivity between the client and the remote data center: quality of network, intermittent network faults in the client's LAN or WAN infrastructure and other transient technical reasons.

What Is the Transient Fault Handling Application Block?

The Transient Fault Handling Application Block makes your application more robust by providing the logic for handling transient faults. It does this in two ways.

> Throttling can occur with Windows Azure storage if your client exceeds the scalability targets. For more information, see "*Windows Azure Storage Abstractions and their Scalability Targets*."

Although transient error codes are documented, determining which exceptions are the result of transient faults for a service requires detailed knowledge of and experience using the service. The block encapsulates this kind of knowledge and experience for you.

First, the block includes logic to identify transient faults for a number of common cloud-based services in the form of detection strategies. These detection strategies contain built-in knowledge that is capable of identifying whether a particular exception is likely to be caused by a transient fault condition.

The block includes detection strategies for the following services:

- SQL Database
- Windows Azure Service Bus
- Windows Azure Storage Service
- Windows Azure Caching Service

The Windows Azure Storage detection strategy works with version 2 of the Windows Azure Storage Client Library and with earlier versions. It automatically detects which version of the Storage Client Library you are using and adjusts its detection strategy accordingly.

However, if you are using version 2 of the Windows Azure Storage Client Library, it is recommended that you use the built-in retry policies in the Windows Azure Storage Client Library in preference to the retry policies for the Windows Azure Storage Service in the Transient Fault Handling Application Block.

The Windows Azure Caching detection strategy works with both Windows Azure Caching and Windows Azure Shared Caching.

The Windows Azure Service Bus detection strategy works with both Windows Azure Service Bus and Service Bus for Windows Server.

This kind of retry logic is also known as "conditional retry" logic.

Second, the application block enables you to define your retry strategies so that you can follow a consistent approach to handling transient faults in your applications. The specific retry strategy you use will depend on several factors; for example, how aggressively you want your application to perform retries, and how the service typically behaves when you perform retries. Some services can further throttle or even block client applications that retry too aggressively. A retry strategy defines how many retries you want to make before you decide that the fault is not transient or that you cannot wait for it to be resolved, and what the intervals should be between the retries.

The built-in retry strategies allow you to specify that retries should happen at fixed intervals, at intervals that increase by the same amount each time, and at intervals that increase exponentially but with some random variation. The following table shows examples of all three strategies.

Retry strategy	**Example (intervals between retries in seconds**
Fixed interval	2,2,2,2,2,2
Incremental intervals	2,4,6,8,10,12
Random exponential back-off intervals	2, 3.755, 9.176, 14.306, 31.895

All retry strategies specify a maximum number of retries after which the exception from the last attempt is allowed to bubble up to your application.

In many cases, you should use the random exponential back-off strategy to gracefully back off the load on the service. This is especially true if the service is throttling client requests.

You can define your own custom detection strategies if the built-in detection strategies included with the application block do not meet your requirements. The application block also allows you to define your own custom retry strategies that define additional patterns for retry intervals.

Figure 1 illustrates how the key elements of the Transient Fault Handling Application Block work together to enable you to add the retry logic to your application.

High throughput applications should typically use an exponential back-off strategy. However, for user-facing applications such as websites you may want to consider a linear back-off strategy to maintain the responsiveness of the UI.

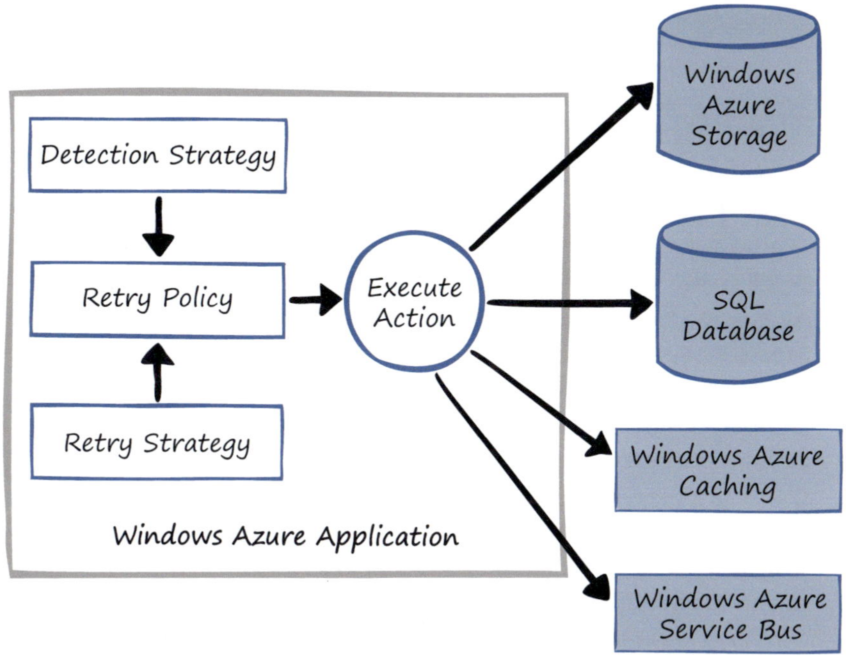

In many cases, retrying immediately may succeed without the need to wait. By default, the block performs the first retry immediately before using the retry intervals defined by the strategy.

FIGURE 1
The Transient Fault Handling Application Block

A retry policy combines a detection strategy with a retry strategy. You can then use one of the overloaded versions of the **ExecuteAction** method to wrap the call that your application makes to one of the services.

You must select the appropriate detection strategy for the service whose method you are calling from your Windows Azure application.

Historical Note

The Transient Fault Handling Application Block is a product of the collaboration between the *Microsoft patterns & practices* team and the *Windows Azure Customer Advisory Team*. It is based on the initial detection and retry strategies, and the data access support from the "*Transient Fault Handling Framework for SQL Database, Windows Azure Storage, Service Bus & Cache*." The new application block now includes enhanced configuration support, enhanced support for wrapping asynchronous calls, and provides integration of the application block's retry strategies with the Windows Azure storage retry mechanism. The Transient Fault Handling Application Block supersedes the Transient Fault Handling Framework and is now the recommended approach to handling transient faults in Windows Azure applications.

Using the Transient Fault Handling Application Block

This section describes, at a high-level, how to use the Transient Fault Handling Application Block. It is divided into the following main subsections. The order of these sections reflects the order in which you would typically perform the associated tasks.

- **Adding the Transient Fault Handling Application Block to your Visual Studio Project**. This section describes how you can prepare your Visual Studio solution to use the block.
- **Defining a retry strategy**. This section describes the ways that you can define a retry strategy in your application.
- **Defining a retry policy**. This section describes how you can define a retry policy in your application.
- **Executing an operation with a retry policy**. This section describes how to execute actions with a retry policy to handle any transient faults.

  *A **retry policy** is the combination of a retry strategy and a detection strategy. You use a retry policy when you execute an operation that may be affected by transient faults.*

NuGet makes it very easy for you to configure your project with all of the prerequisites for using the Transient Fault Handling Application Block.

For detailed information about configuring the Transient Fault Handling Application Block and writing code that uses the Transient Fault Handling Application Block, see the topic "*The Transient Fault Handling Application Block*" in the Enterprise Library Reference Documentation.

Adding the Transient Fault Handling Application Block to Your Visual Studio Project

As a developer, before you can write any code that uses the Transient Fault Handling Application Block, you must configure your Visual Studio project with all of the necessary assemblies, references, and other resources that you'll need.

Instantiating the Transient Fault Handling Application Block Objects

There are two basic approaches to instantiating the objects from the application block that your application requires. In the first approach, you can explicitly instantiate all the objects in code, as shown in the following code snippet:

```
var retryStrategy = new Incremental(5, TimeSpan.FromSeconds(1), TimeSpan.FromSeconds(2));
var retryPolicy =
    new RetryPolicy<SqlDatabaseTransientErrorDetectionStrategy>(retryStrategy);
```

*If you instantiate the **RetryPolicy** object using **new**, you cannot use the default strategies defined in the configuration.*

In the second approach, you can instantiate and configure the objects from configuration data as shown in the following code snippet:

```
// Load policies from the configuration file.
// SystemConfigurationSource is defined in
// Microsoft.Practices.EnterpriseLibrary.Common.
using (var config = new SystemConfigurationSource())
{
  var settings = RetryPolicyConfigurationSettings.GetRetryPolicySettings(config);

  // Initialize the RetryPolicyFactory with a RetryManager built from the
  // settings in the configuration file.
  RetryPolicyFactory.SetRetryManager(settings.BuildRetryManager());

  var retryPolicy = RetryPolicyFactory.GetRetryPolicy
                  <SqlDatabaseTransientErrorDetectionStrategy>("Incremental Retry Strategy");
   ...
   // Use the policy to handle the retries of an operation.
}
```

Defining a Retry Strategy

There are three considerations in defining retry strategies for your application: which retry strategy to use, where to define the retry strategy, and whether to use default retry strategies.

In most cases, you should use one of the built-in retry strategies: fixed interval, incremental, or random exponential back off. You configure each of these strategies using custom sets of parameters to meet your application's requirements; the parameters specify when the strategy should stop retrying an operation, and what the intervals between the retries should be. The choice of retry strategy will be largely determined by the specific requirements of your application and the characteristics of the service invoked. For more details about the parameters for each retry strategy, see the topic "*Source Schema for the Transient Fault Handling Application Block*" in the Enterprise Library Reference Documentation.

You can define your own custom retry strategy. For more information, see the topic "*Implementing a Custom Retry Strategy*."

You can define your retry policies either in code or in the application configuration file. Defining your retry policies in code is most appropriate for small applications with a limited number of calls that require retry logic. Defining the retry policies in configuration is more useful if you have a large number of operations that require retry logic, because it makes it easier to maintain and modify the policies.

For more information about how to define your retry strategy in code, see the topic "*Specifying Retry Strategies in Code*."

For more information about how to define your retry strategies in a configuration file, see the topic "*Specifying Retry Strategies in the Configuration*."

If you define your retry strategies in the configuration file for the application, you can also define default retry strategies. The block allows you to specify default retry strategies at two levels. You can specify a default retry strategy for each of the following operation categories: SQL connection operations, SQL command operations, Windows Azure Service Bus operations, Windows Azure Caching, and Windows Azure Storage operations. You can also specify a global default retry strategy. For more information, see the topic "*Entering Configuration Information*."

Defining a Retry Policy

If you are using Windows Azure Storage and you are already using the retry policies mechanism in the **Microsoft.WindowsAzure.StorageClient** namespace, then you can use retry strategies from the application block and configure the Windows Azure Storage client API to take advantage of the extensible retry functionality provided by the application block.

A retry policy is the combination of a retry strategy and a detection strategy that you use when you execute an operation that may be affected by transient faults. The **RetryManager** class includes methods that enable you to create retry policies by explicitly identifying the retry strategy and detection strategy, or by using default retry strategies defined in the configuration file.

For more information about using the retry policies, see the topic "*Key Scenarios*" on MSDN.

For more information about the **RetryPolicy** delegate in the **Microsoft.WindowsAzure.StorageClient** namespace, see the blog post "*Overview of Retry Policies in the Windows Azure Storage Client Library.*"

Executing an Operation with a Retry Policy

The **RetryPolicy** class includes several overloaded versions of the **ExecuteAction** method. You use the **ExecuteAction** method to wrap the synchronous calls in your application that may be affected by transient faults. The different overloaded versions enable you to wrap the following types of calls to a service.

- Synchronous calls that return a **void**.
- Synchronous calls that return a value.

The **RetryPolicy** class includes several overloaded versions of the **ExecuteAsync** method. You use the **ExecuteAsync** method to wrap the asynchronous calls in your application that may be affected by transient faults. The different overloaded versions enable you to wrap the following types of calls to a service.

- Asynchronous calls that return a **Task**.
- Asynchronous calls that return a **Task<T>**.

There are also overloaded versions of the **ExecuteAsync** method that include a cancellation token parameter that enables you to cancel the retry operations on the asynchronous methods after you have invoked them.

The **ExecuteAction** and **ExecuteAsync** methods automatically apply the configured retry strategy and detection strategy when they invoke the specified action. If no transient fault manifests itself during the invocation, your application continues as normal, as if there was nothing between your code and the action being invoked. If a transient fault does manifest itself, the block will initiate the recovery by attempting to invoke the specified action multiple times as defined in the retry strategy. As soon as a retry attempt succeeds, your application continues as normal. If the block does not succeed in executing the operation within the number of retries specified by the retry strategy, then the block rethrows the exception to your application. Your application must still handle this exception properly. Sometimes, the operation to retry might be more than just a single method in your object. For example, if you are trying to send a message using Service Bus, you cannot try to resend a failed brokered message in a retry; you must create a new message for each retry and ensure that messages are properly disposed. Typically, you would wrap this behavior in your own method to use with the block, but this may affect the design of your API, especially if you have a component that sends messages with a retry policy on behalf of other code in your application.

The Transient Fault Handling Application Block is not a substitute for proper exception handling. Your application must still handle any exceptions that are thrown by the service you are using. You should consider using the Exception Handling Application Block described in Chapter 3, "Error Management Made Exceptionally Easy."

You can use the **Retrying** event to receive notifications in your application about the retry operations that the block performs.

If you are working with SQL Database, the application block includes classes that provide direct support for SQL Database, such as the **ReliableSqlConnection** class. These classes will help you reduce the amount of code you need to write. However, you are then using a specific class from the block instead of the standard ADO.NET class. Also, these classes don't work with Entity Framework.

In addition, the application block includes classes that wrap many common SQL Database operations with a retry policy for you. Using these classes minimizes the amount of code you need to write.

For more information about executing an operation with a retry policy, see the topic "*Key Scenarios*" in the Enterprise Library Reference Documentation.

When Should You Use the Transient Fault Handling Application Block?

This section describes two scenarios in which you should consider using the Transient Fault Handling Application Block in your Windows Azure solution.

You are Using a Windows Azure Service

If your application uses any of the Windows Azure services supported by the Transient Fault Handling Application Block (SQL Database, Windows Azure Storage, Windows Azure Caching, or Windows Azure Service Bus), then you can make your application more robust by using the application block. Additionally, by using the block you will be following the documented, recommended practices for these services: published information about the error codes from these services is prescriptive and indicates how you should attempt to retry operations. Any Windows Azure application that uses these services may occasionally encounter transient faults with these services. Although you could add your own detection logic to your application, the application block's built-in detection strategies will handle a wider range of transient faults. It is also quicker and easier to use the application block instead of developing your own solution; this is especially true of asynchronous operations, which can appear to be complex.

The Windows Azure Storage Client Library version 2 includes retry policies for Windows Azure Storage Services and it is recommended that you use these built-in policies in preference to the Transient Fault Handling Application Block.

For more information about retries in Windows Azure storage, see "*Overview of Retry Policies in the Windows Azure Storage Client Library*."

You are Using Service Bus for Window Server

If you are using Service Bus for Windows Server in your solution, you can use the Windows Azure Server Bus detection to detect any transient faults when you use the service bus. You can use the Windows Azure Service Bus detection strategy with Service Bus for Windows Server in exactly the same way that you use it with Windows Azure Service Bus.

You Are Using a Custom Service

If your application uses a custom service, it can still benefit from using the Transient Fault Handling Application Block. You can author a custom detection strategy for your service that encapsulates your knowledge of which transient exceptions may result from a service invocation. The Transient Fault Handling Application Block then provides you with the framework for defining retry policies and for wrapping your method calls so that the application block applies the retry logic.

More Information

All links in this book are accessible from the book's online bibliography on MSDN at *http://aka.ms/el6biblio*.

For detailed information about configuring the Transient Fault Handling Application Block and writing code that uses the Transient Fault Handling Application Block, see the topic "*The Transient Fault Handling Application Block*" in the Enterprise Library Reference Documentation.

For information about how you can use NuGet to prepare your Visual Studio project to work with the Transient Fault Handling Application Block, see the topic "*Adding the Transient Fault Handling Application Block to Your Solution*."

For more information about throttling in Windows Azure, see "*Windows Azure Storage Abstractions and their Scalability Targets*" on MSDN.

There is an additional approach that is provided for backward compatibility with the "*Transient Fault Handling Application Framework*" that uses the **RetryPolicyFactory** class.

For more details about the parameters for each retry strategy, see the topic "*Source Schema for the Transient Fault Handling Application Block*" in the Enterprise Library Reference Documentation.

You can define your own custom retry strategy. For more information, see the topic "*Implementing a Custom Retry Strategy*."

For more information about how to define your retry strategy in code, see the topic "*Specifying Retry Strategies in Code*."

For more information about how to define your retry strategies in a configuration file, see the topic "*Specifying Retry Strategies in the Configuration*."

If you define your retry strategies in the configuration file for the application, you can also define default retry strategies and a global default retry strategy. For more information, see the topic "*Entering Configuration Information*."

For more information about using the retry policies, see the topic "*Key Scenarios*" on MSDN.

For more information about the **RetryPolicy** delegate in the **Microsoft.WindowsAzure.StorageClient** namespace, see the blog post "*Overview of Retry Policies in the Windows Azure Storage Client Library*."

For more information about retries in Windows Azure storage, see "*Overview of Retry Policies in the Windows Azure Storage Client Library*" and "*Windows Azure Storage Client Library 2.0 Breaking Changes & Migration Guide*."

The Transient Fault Handling Application Block is a product of the collaboration between the Microsoft *patterns & practices team* and the *Windows Azure Customer Advisory Team*. It is based on the initial detection and retry strategies, and the data access support from the "*Transient Fault Handling Framework for SQL Database, Windows Azure Storage, Service Bus & Cache*" on MSDN.

General Links:

- There are resources to help if you're getting started with Enterprise Library, and there's help for existing users as well (such as the breaking changes and migration information for previous versions) available on the Enterprise Library Community Site at *http://www.codeplex.com/entlib/*.
- For more details about the Enterprise Library Application Blocks, see the *Enterprise Library Reference Documentation* and the *Enterprise Library 6 Class Library*.
- You can download and work through the Hands-On Labs for Enterprise Library, which are available at *http://aka.ms/el6hols*.
- You can download the Enterprise Library binaries, source code, Reference Implementation, or Quick-Starts from the Microsoft Download Center at *http://go.microsoft.com/fwlink/p/?LinkID=290898*.
- To help you understand how you can use Enterprise Library application blocks, we provide a series of simple example applications that you can run and examine. To obtain the example applications, go to *http://go.microsoft.com/fwlink/p/?LinkID=304210*.

5 As Easy As Falling Off a Log
Using the Logging Application Block

Introduction

Just in case you didn't quite grasp it from the title, this chapter is about one of the most useful and popular of the Enterprise Library blocks, the Logging Application Block, which makes it really easy to perform logging in a myriad of different ways depending on the requirements of your application.

Logging generally fulfills two main requirements: monitoring general application performance, and providing information. In terms of performance, logging allows you to monitor what's happening inside your application and, in some cases, what's happening in the world outside as well. For example, logging can indicate what errors or failures have occurred, when something that should have happened did not, and when things are taking a lot longer than they should. It can also simply provide status information on processes that are working correctly—including those that talk to the outside world. Let's face it, there's nothing more rewarding for an administrator than seeing an event log full of those nice blue information icons.

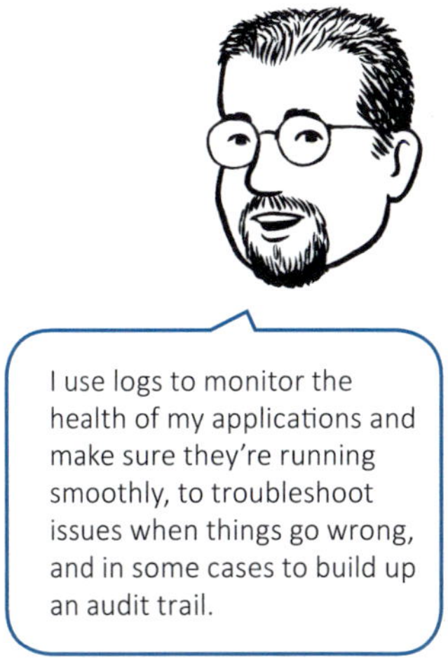

Secondly, and possibly even more importantly, logging can provide vital information about your application. Often referred to as auditing, this type of logging allows you to track the behavior of users and processes in terms of the tasks they carry out, the information they read and change, and the resources they access. It can provide an audit trail that allows you to follow up and get information about malicious activity (whether it succeeds or not), will allow you to trace events that may indicate future attack vectors or reveal security weaknesses, and even help you to recover when disaster strikes (though this doesn't mean you shouldn't be taking the usual precautions such as backing up systems and data). One other area where audit logging is useful is in managing repudiation. For example, your audit logs may be useful in legal or procedural situations where users or external attackers deny their actions.

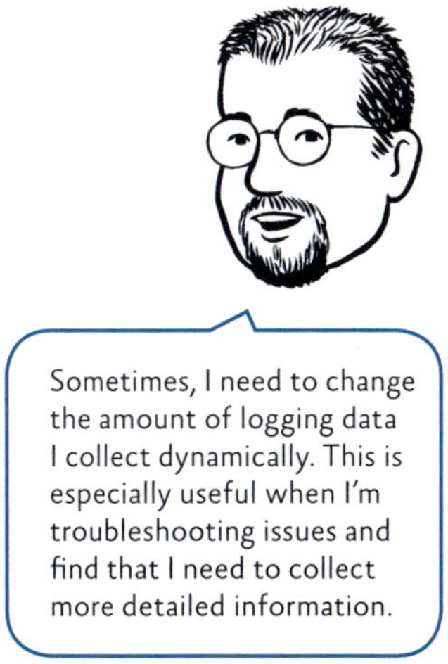

The Logging block is a highly flexible and configurable solution that allows you to create and store log messages in a wide variety of locations, categorize and filter messages, and collect contextual information useful for debugging and tracing as well as for auditing and general logging requirements. It abstracts the logging functionality from the log destination so that the application code is consistent, irrespective of the location and type of the target logging store. Changes to almost all of the parameters that control logging are possible simply by changing the configuration after deployment and at run time. This means that administrators and operators can vary the logging behavior as they manage the application.

If you are concerned about the potential overhead of logging on the performance of your application, or you want to use more structured log messages, or you plan to automate processes from log entries, you should consider using the Semantic Logging Application Block instead of the Logging Application Block. The Semantic Logging Application Block is described in the next chapter.

What Does the Logging Block Do?

The Logging Application Block allows you to decouple your logging functionality from your application code. The block can route log entries to a Windows Event Log, a database, or a text (or XML) file. It can also generate an email message containing the logging information, a message you can route through Windows Message Queuing (using a distributor service provided with the block). And, if none of these built-in capabilities meets your requirements, you can create a provider that sends the log entry to any other custom location or executes some other action.

In your application, you simply generate a log entry using a suitable logging object, such as the **LogWriter** class, and then call a method to write the information it contains to the logging system. The Logging block routes the log message through any filters you define in your configuration, and on to the listeners that you configure. Each listener defines the target of the log entry, such as Windows Event Log or an email message, and uses a formatter to generate suitably formatted content for that logging target.

You can see from this that there are many objects involved in this multi-step process, and it is important to understand how they interact and how the log message flows through the pipeline of processes. Figure 1 shows the overall process in more detail, and provides an explanation of each stage.

Add code to generate log entries throughout your application wherever you think that logging information may prove useful in the future. You can use configuration to control how much and where information is logged.

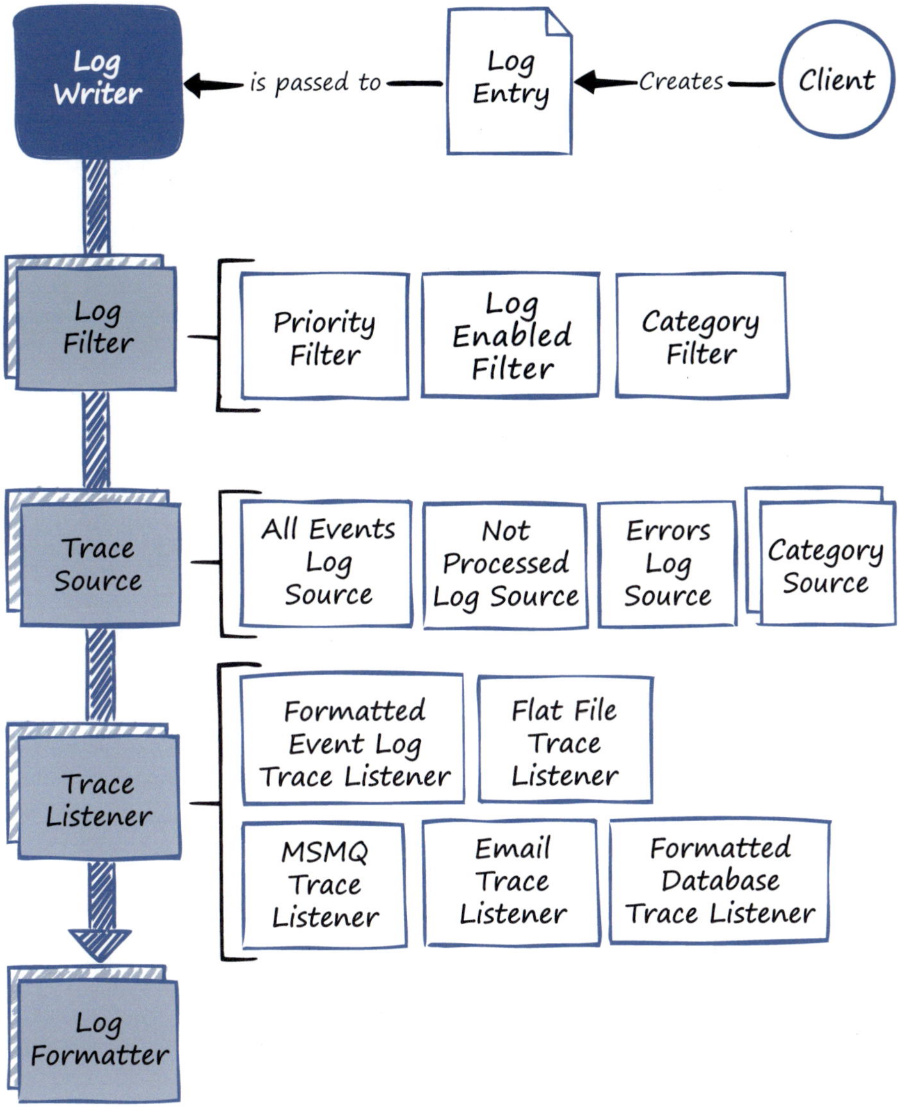

Figure 1
An overview of the logging process and the objects in the Logging block

Stage	Description
Creating the Log Entry	The user creates a **LogWriter** instance, and invokes one of the **Write** method overloads. Alternatively, the user can create a new **LogEntry** explicitly, populate it with the required information, and use a **LogWriter** to pass it to the Logging block for processing.
Filtering the Log Entry	The Logging block filters the **LogEntry** (based on your configuration settings) for message priority, or categories you added to the **LogEntry** when you created it. It also checks to see if logging is enabled. These filters can prevent any further processing of the log entries. This is useful, for example, when you want to allow administrators to enable and disable additional debug information logging without requiring them to restart the application.
Selecting Trace Sources	Trace sources act as the link between the log entries and the log targets. There is a trace source for each category you define in the logging block configuration; plus, there are three built-in trace sources that capture all log entries, unprocessed entries that do not match any category, and entries that cannot be processed due to an error while logging (such as an error while writing to the target log).
Selecting Trace Listeners	Each trace source has one or more trace listeners defined. These listeners are responsible for taking the log entry, passing it through a separate log formatter that translates the content into a suitable format, and passing it to the target log. Several trace listeners are provided with the block, and you can create your own if required.
Formatting the Log Entry	Each trace listener can use a log formatter to format the information contained in the log entry. The block contains log message formatters, and you can create your own formatter if required. The text formatter uses a template containing placeholders that makes it easy to generate the required format for log entries.

The developers and IT Pros should discuss which categories will be useful when configuring the block.

Logging Categories

Categories allow you to specify the target(s) for log entries processed by the block. You can define categories that relate to one or more targets. For example, you might create a category named General containing trace listeners that write to text files and XML files, and a category named Auditing for administrative information that is configured to use trace listeners that write to one or more databases. Then you can assign a log entry to one or more categories, effectively mapping it to multiple targets. The three log sources shown in the schematic in Figure 1 (all events log source, not processed log source, and errors log source) are themselves categories for which you can define trace listeners.

> *Logging is an added-value service for applications, and so any failures in the logging process must be handled gracefully without raising an exception to the main business processes. The Logging block achieves this by sending all logging failures to a special category (the errors log source) which is named **Logging Errors & Warnings**. By default, these error messages are written to Windows Event Log, though you can configure this category to write to other targets using different trace listeners if you wish.*

If minimizing the overhead of logging is a significant concern, you should consider logging asynchronously as described later in this chapter. You should also consider using the Semantic Logging Application Block in its out-of-process mode; this is described in the next chapter.

Logging Overhead and Additional Context Information

No matter how you implement logging, it will always have some performance impact. The Logging block provides a flexible and highly configurable logging solution that is carefully designed to minimize performance impact. However, you should be aware of this impact, and consider how your own logging strategy will affect it. For example, a complex configuration that writes log entries to multiple logs and uses multiple filters is likely to have more impact than simple configurations. You must balance your requirements for logging against performance and scalability needs.

The **LogWriter** automatically collects some context information such as the time, the application domain, the machine name, and the process ID. To minimize the impact on performance, the **LogWriter** class caches some of these values and uses lazy initialization so the values are only collected if they are actually retrieved by a listener.

However, collecting additional context information can be expensive in processing terms and, if you are not going to use the information, wastes precious resources and may affect performance. Therefore, the Logging block only collects other less commonly used information from the environment, which you might require only occasionally, if you specify that you want this information when you create the LogEntry instance. Four classes within the Logging block can collect specific sets of context information that you can add to your log entry. This includes COM+ diagnostic information, the current stack trace, the security-related information from the managed run-time, and security-related information from the operating system. There is also a dictionary property for the log entry where you can add any additional custom information you require, and which must appear in your logs.

How Do I Use the Logging Block?

It's time to see some examples of the Logging block use, including how to create log entries and write them to various targets such as the Windows Event Log, disk files, and a database. Later you'll see how you can use some of the advanced features of the block, such as checking filter status and adding context information to a log entry. However, before you can start using the Logging block, you must configure it.

Adding the Logging Block to Your Project

Before you start editing your code to use the Logging block, you must add the block to your project. You can use the NuGet Package Manager to add the required assemblies and references. In the **Manage Nuget Packages** dialog in Visual Studio, search online for the **EnterpriseLibrary.Logging** package and install it.

If you intend to send log entries to a database, you must also install the **EnterpriseLibrary.Logging.Database** package. If you intend to use remote logging, you must also install the **EnterpriseLibrary.Logging.Service** package on the system where you are collecting the logs.

Now you are ready to add some configuration and write some code.

Configuring the Logging Block

You can configure the Logging block using the programmatic configuration approach described in Chapter 1, "Introduction to the Microsoft Enterprise Library." The Logging block includes a **LoggingConfiguration** class that you can use to build the configuration data that you need. Typically, you create the formatters and trace listeners that you need, and then add and the log sources to the configuration. The following code sample shows how to create a **LoggingConfiguration** instance and then use it to create a **LogWriter**.

```
// Formatter
TextFormatter briefFormatter = new TextFormatter(...);

// Trace Listener
var flatFileTraceListener = new FlatFileTraceListener(
                                @"C:\Temp\FlatFile.log",
                                "----------------------------------------",
                                "----------------------------------------",
                                briefFormatter);

// Build Configuration
var config = new LoggingConfiguration();
config.AddLogSource("DiskFiles", SourceLevels.All, true)
    .AddTraceListener(flatFileTraceListener);
LogWriter defaultWriter = new LogWriter(config);
```

This is a simplified version of the configuration that the sample in this chapter uses, you can see the complete configuration in the method **BuildProgrammaticConfig**. If you examine this method, you'll also see how to configure filters and the special sources.

You should dispose the **LogWriter** instance when you have finished logging in your application to ensure that all resources are released.

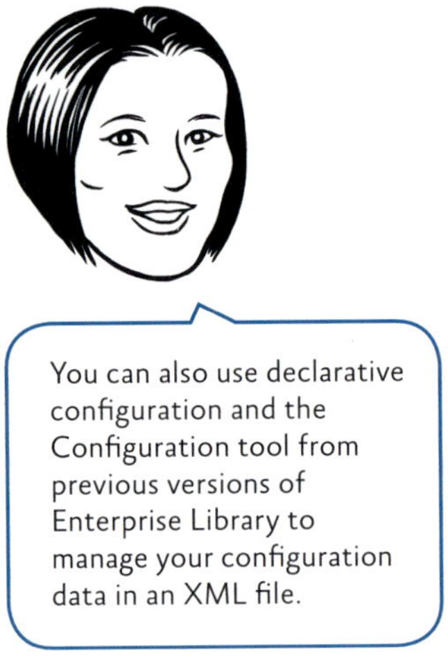

If you prefer to use declarative configuration, you can use the **LogWriterFactory** class to load the configuration data. The sample application shows how to use this approach to load configuration data from the app.config file.

```
LogWriterFactory logWriterFactory = new LogWriterFactory();
LogWriter logWriter = logWriterFactory.Create();
```

Diving in with an Example

To demonstrate the features of the Logging block, we provide a *sample application* that you can download and run on your own computer. You can run the executable directly from the bin\Debug folder, or you can open the solution named **Logging** in Visual Studio to see the code and run it under Visual Studio. The application includes a preconfigured database for storing log entries, as well as scripts you can use to create the Logging database within a different database server if you prefer.

> *You do not need to run the scripts if you have Microsoft SQL Server LocalDB installed locally. If you want to specify a different database for logging, edit the script named CreateLoggingDb.cmd to specify the location of the database and execute it. After you do that, you must change the connection string named* ***ExampleDatabase*** *to point to your new database.*
>
> *In addition, depending on the version of the operating system you are using, you may need to execute the application under the context of an account with administrative privileges. If you are running the sample from within Visual Studio, start Visual Studio by right-clicking the entry in your Start menu and selecting* ***Run as administrator****.*

One other point to note about the sample application is that it creates a folder named **Temp** in the root of your C: drive if one does not already exist, and writes the text log files there so that you can easily find and view them.

Creating and Writing Log Entries with a LogWriter

The first of the examples, *Simple logging with the Write method of a LogWriter*, demonstrates how you can use a **LogWriter** directly to create log entries. The first stage is to obtain a **LogWriter** instance, and the example uses the simplest approach—passing a **LoggingConfiguration** object to the **LogWriter** constructor. See the following code.

```
LoggingConfiguration loggingConfiguration = BuildProgrammaticConfig();
LogWriter defaultWriter = new LogWriter(loggingConfiguration);
```

Now you can call the Write method and pass in any parameter values you require. There are many overloads of the Write method. They allow you to specify the message text, the category, the priority (a numeric value), the event ID, the severity (a value from the TraceEventType enumeration), and a title for the event. There is also an overload that allows you to add custom values to the log entry by populating a Dictionary with name and value pairs (you will see this used in a later example). Our example code uses several of these overloads. We've removed some of the Console.WriteLine statements from the code listed here to make it easier to see what it actually does.

```
// Check if logging is enabled before creating log entries.
if (defaultWriter.IsLoggingEnabled())
{
  defaultWriter.Write("Log entry created using the simplest overload.");
  defaultWriter.Write("Log entry with a single category.", "General");
  defaultWriter.Write("Log entry with a category, priority, and event ID.",
                      "General", 6, 9001);
  defaultWriter.Write("Log entry with a category, priority, event ID, "
                      + "and severity.", "General", 5, 9002,
                      TraceEventType.Warning);
  defaultWriter.Write("Log entry with a category, priority, event ID, "
                      + "severity, and title.", "General", 8, 9003,
                      TraceEventType.Warning, "Logging Block Examples");
}
else
{
  Console.WriteLine("Logging is disabled in the configuration.");
}
```

Notice how the code first checks to see if logging is enabled. There is no point using valuable processor cycles and memory generating log entries if they aren't going anywhere. If you add a **LogEnabledFilter** instance to the **LoggingConfiguration** object, you can use it to control logging for the whole block (you can see this in the example application). This filter has the single property, **Enabled**, that allows administrators to enable and disable all logging for the block. When it is set to False, the **IsLoggingEnabled** property of the **LogWriter** will return **false** as well.

The example produces the following result. All of the events are sent to the **General** category, which is configured to write events to the Windows Application Event Log (this is the default configuration for the block).

```
Created a Log Entry using the simplest overload.
Created a Log Entry with a single category.
Created a Log Entry with a category, priority, and event ID.
Created a Log Entry with a category, priority, event ID, and severity.
Created a Log Entry with a category, priority, event ID, severity, and title.
Open Windows Event Viewer 'Application' Log to see the results.
```

You can open Windows Event Viewer to see the results. Figure 2 shows the event generated by the last of the **Write** statements in this example.

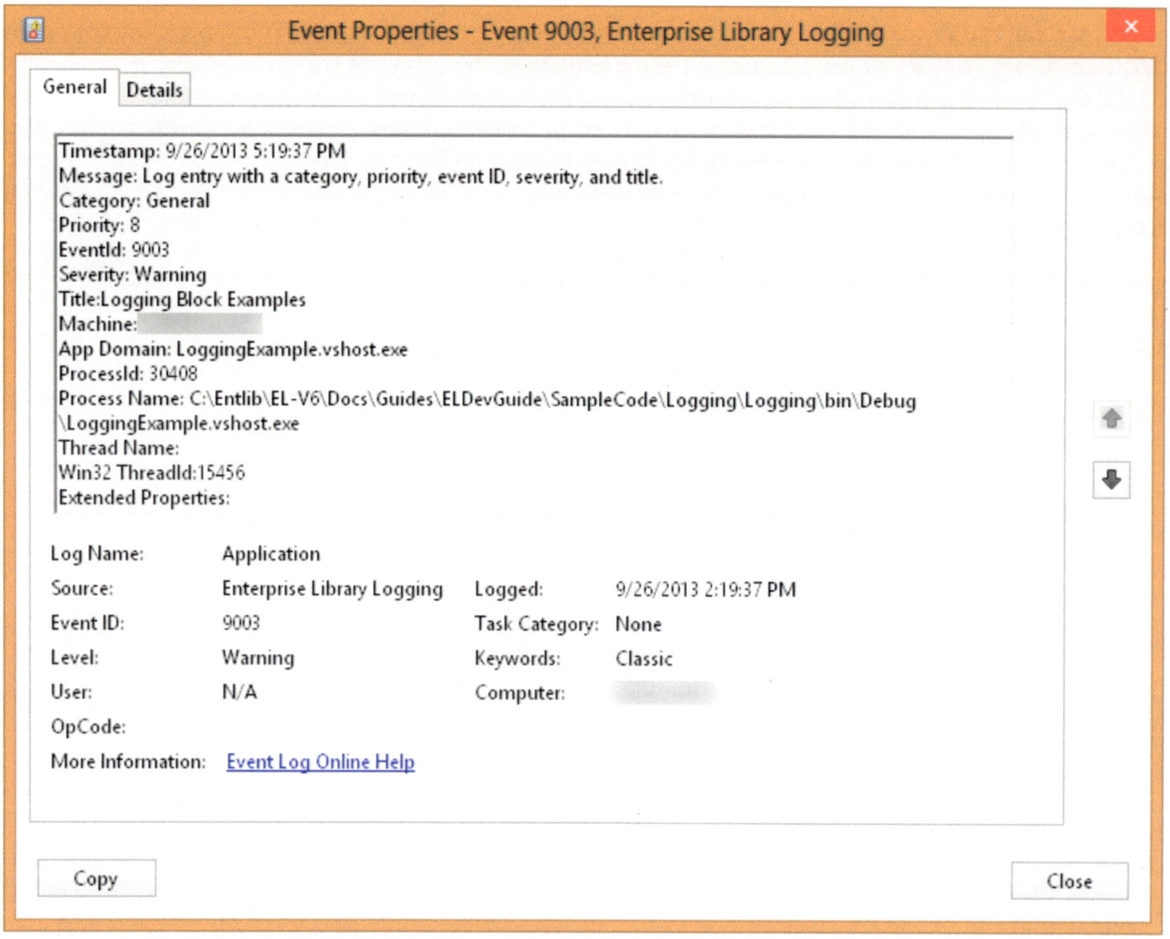

FIGURE 2
The logged event

> *If you do not specify a value for one of the parameters of the **Write** method, the Logging block uses the default value for that parameter. The defaults are Category = General, Priority = -1, Event ID = 1, Severity = Information, and an empty string for Title.*

About Logging Categories

Categories are the way that Enterprise Library routes events sent to the block to the appropriate target, such as a database, the event log, an e-mail message, and more. The previous example makes use of the default configuration for the Logging block. When you add the Logging block to your application configuration using the Enterprise Library configuration tools, it contains the single category named **General** that is configured to write events to the Windows Application Event Log.

You can change the behavior of logging for any category. For example, you can change the behavior of the previous example by reconfiguring the event log trace listener specified for the **General** category, or by reconfiguring the text formatter that this trace listener uses. You can change the event log to which the event log trace listener sends events; edit the template used by the text formatter; or add other trace listeners.

However, it's likely that your application will need to perform different types of logging for different tasks. The typical way to achieve this is to define additional categories, and then specify the type of trace listener you need for each category. For example, you may want to send audit information to a text file or an XML file, to a database, or both; instead of to Windows Event Log. Alternatively, you may want to send indications of catastrophic failures to administrators as email messages; in this scenario, the severity of the event rather than the category determines where to send it. If you are using an enterprise-level monitoring system, you may instead prefer to send events to another system through Windows Message Queuing.

You can easily add categories to your application configuration. The approach is to create the trace listeners for the logging targets you require, such as the flat file trace listener or database trace listener, and then use the **AddLogSource** method to add the categories you require.

The following code sample shows how you can configure the General category to output log messages to the Windows Event Log and the Important category to output log messages to the event log and a rolling flat file.

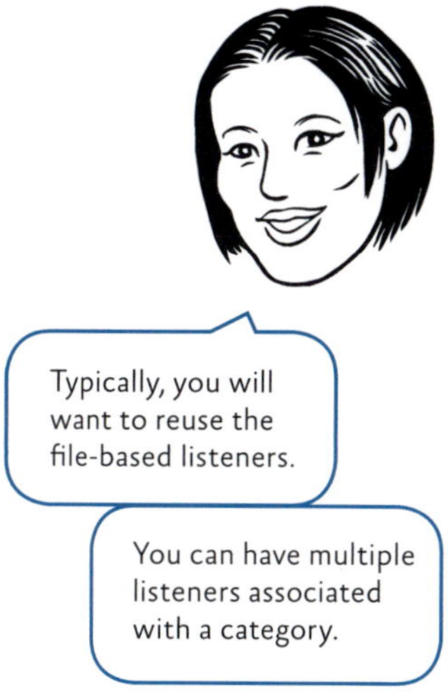


```
var rollingFlatFileTraceListener =
  new RollingFlatFileTraceListener(@"C:\Logs\RollingFlatFile.log",
      "----------------------------------------",
      "----------------------------------------",
      extendedFormatter, 20, "yyyy-MM-dd",
      RollFileExistsBehavior.Increment,RollInterval.None, 3);
var eventLog = new EventLog("Application", ".", "Enterprise Library Logging");
var eventLogTraceListener = new FormattedEventLogTraceListener(eventLog);

config.AddLogSource("General", SourceLevels.All, true)
      .AddTraceListener(eventLogTraceListener);
config.AddLogSource("Important", SourceLevels.All, true)
      .AddTraceListener(eventLogTraceListener);
config.LogSources["Important"]
      .AddTraceListener(rollingFlatFileTraceListener);
```

You can specify two properties for each category (source) you add. You can specify a filtering level for the log source using the **SourceLevels** enumeration. You can set the **AutoFlush** property to specify that the block should flush log entries to their configured target trace listeners each time as soon as they are written to the block, or only when you call the **Flush** method of the listener. If you set the **AutoFlush** property to **False**, ensure that your code calls this method when an exception or failure occurs to avoid losing any cached logging information.

The other property you can set for each category is the **SourceLevels** property (which sets the **Source Levels** property of each listener). This specifies the minimum severity (such as Warning or Critical) for the log entries that the category filter will pass to its configured trace listeners. Any log entries with a lower severity will be blocked. The default severity is **All**, and so no log entries will be blocked unless you change this value. You can also configure a **Severity Filter** (which sets the **Filter** property) for each individual trace listener, and these values can be different for trace listeners in the same category. You will see how to use the **Filter** property of a trace listener in the next example in this chapter.

Filtering by Category

You can use the **CategoryFilter** class to define a filter that you can use to filter log entries sent to the block based on their membership in specified categories. You can add multiple categories to your configuration to manage filtering, though overuse of this capability can make it difficult to manage logging.

To define your category filters, create a list of category names to pass the constructor of the **CategoryFilter** class. The constructor also has a parameter that enables you to specify the filter mode (Allow all except..., or Deny all except...). The example application contains only a single filter that is configured to allow logging to all categories except for the category named (rather appropriately) **BlockedByFilter**. You will see the **BlockedByFilter** category used in the section "Capturing Unprocessed Events and Logging Errors" later in this chapter.

Writing Log Entries to Multiple Categories

In addition to being able to define multiple categories, you can send a log entry to more than one category in a single operation. This approach often means you can define fewer categories, and it simplifies the configuration because each category can focus on a specific task. You don't need to have multiple categories with similar sets of trace listeners.

The second example, *Logging to multiple categories with the Write method of a LogWriter*, shows how to write to multiple categories. The example has two categories, named **DiskFiles** and **Important**, defined in the configuration. The **DiskFiles** category contains references to a flat file trace listener and an XML trace listener. The **Important** category contains references to an event log trace listener and a rolling flat file trace listener.

The example uses the following code to create an array of the two category names, **DiskFiles** and **Important**, and then it writes three log messages to these two categories using the **Write** method of the **LogWriter** in the same way as in the previous example. Again, we've removed some of the **Console.WriteLine** statements to make it easier to see what the code actually does.

```
// Check if logging is enabled before creating log entries.
if (defaultWriter.IsLoggingEnabled())
{
  // Create a string array (or List<>) containing the categories.
  string[] logCategories = new string[] {"DiskFiles", "Important"};
  // Write the log entries using these categories.
  defaultWriter.Write("Log entry with multiple categories.", logCategories);
  defaultWriter.Write("Log entry with multiple categories, a priority, "
                    + "and an event ID.", logCategories, 7, 9004);
  defaultWriter.Write("Log entry with multiple categories, a priority, "
                    + "event ID, severity, and title.", logCategories, 10,
                    9005, TraceEventType.Critical, "Logging Block Examples");
}
else
{
  Console.WriteLine("Logging is disabled in the configuration.");
}
```

Controlling Output Formatting

If you run the example above and then open Windows Event Log, you will see the three events generated by this example. Also, in the C:\Temp folder, you will see three files. RollingFlatFile.log is generated by the rolling flat file trace listener, and contains the same information as the event log event generated by the event log trace listener. If you explore the configuration, you will see that they both use the same text formatter to format the output.

The FlatFile.log file, which is generated by the flat file trace listener, contains only a simplified set of values for each event. For example, this is the output generated for the last of the three log entries.

```
----------------------------------------
Timestamp: 24/11/2009 10:49:26
Message: Log entry with multiple categories, a priority, event ID, severity, and title.
Category: DiskFiles, Important
Priority: 10
EventId: 9005
ActivityId: 00000000-0000-0000-0000-000000000000
Severity: Critical
Title:Logging Block Examples
----------------------------------------
```

The reason is that the flat file trace listener is configured to use a different text formatter—in this case one named **briefFormatter**. Most trace listeners use a formatter to translate the contents of the log entry properties into the appropriate format for the target of that trace listener. Trace listeners that create text output, such as a text file or an e-mail message, use a text formatter defined within the configuration of the block. You can also format log entries using XML or JSON formatters.

If you examine the configured text formatter, you will see that it has a **Template** property. A full list of tokens and their meaning is available in the online documentation for Enterprise Library, although most are fairly self-explanatory.

> *If you use the **XmlLogFormatter** class to save the log messages formatted as XML in a flat file, you will need to perform some post-processing on the log files before you can use a tool such as Log Parser to read them. This is because the flat file does not have an XML root element, therefore the file does not contain well-formed XML. It's easy to add opening and closing tags for a root element to the start and end of the file.*

Non-Formatted Trace Listeners

While we are discussing output formatting, there is one other factor to consider. Some trace listeners do not use a text formatter to format the output they generate. This is generally because the output is in a binary or specific format. The XML trace listener is a typical example that does not use a text formatter.

For such trace listeners, you can set the **TraceOutputOptions** property to one of a range of values to specify the values you want to include in the output. The **TraceOutputOptions** property accepts a value from the **System.Diagnostics.TraceOptions** enumeration. Valid values include **CallStack**, **DateTime**, **ProcessId**, **LogicalOperationStack**, **Timestamp**, and **ThreadId**. The documentation installed with Enterprise Library, and the documentation for the System.Diagnostics namespace on MSDN, provide more information.

Filtering by Severity in a Trace Listener

The previous example generates a third disk file that we haven't looked at yet. We didn't forget this, but saved it for this section because it demonstrates another feature of the trace listeners that you will often find extremely useful. To see this, you need to view the file XmlLogFile.xml that was generated in the C:\Temp folder by the XML trace listener we used in the previous example. You should open it in Microsoft Internet Explorer (or another web browser or text editor) to see the structure.

You will see that the file contains only one event from the previous example, not the three that the code in the example generated. This is because the XML trace listener has the **Filter** property in its configuration set to **Error**. Therefore, it will log only events with a severity of **Error** or higher. If you look back at the example code, you will see that only the last of the three calls to the **Write** method specified a value for the severity (**TraceEventType.Critical** in this case), and so the default value **Information** was used for the other two events.

*If you get an error indicating that the XML document created by the XML trace listener is invalid, it's probably because you have more than one log entry in the file. This means that it is not a valid XML document—it contains separate event log entries added to the file each time you ran this example. To view it as XML, you must open the file in a text editor and add an opening and closing element (such as **<root>** and **</root>**) around the content. Or, just delete it and run the example once more.*

All trace listeners expose the **Filter** property, and you can use this to limit the log entries written to the logging target to only those that are important to you. If your code generates many information events that you use for monitoring and debugging only under specific circumstances, you can filter these to reduce the growth and size of the log when they are not required.

Alternatively, (as in the example) you can use the **Filter** property to differentiate the granularity of logging for different listeners in the same category. It may be that a flat file trace listener will log all entries to an audit log file for some particular event, but an Email trace listener in the same category will send email messages to administrators only when an Error or Critical event occurs.

Filtering All Log Entries by Priority

As well as being able to filter log entries in individual trace listeners based on their severity, you can set the Logging block to filter all log entries sent to it based on their priority. Alongside the **LogEnabledFilter** class and **CategoryFilter** class (which we discussed earlier in this chapter), you can create **PriorityFilter** instance.

This filter has two properties that you can set: **Minimum Priority** and **Maximum Priority**. The default setting for the priority of a log entry is **-1**, which is the same as the default setting of the **Minimum Priority** property of the filter, and there is no maximum priority set. Therefore, this filter will not block any log entries. However, if you change the defaults for these properties, only log entries with a priority between the configured values (including the specified maximum and minimum values) will be logged. The exception is log entries that have the default priority of **-1**. These are never filtered.

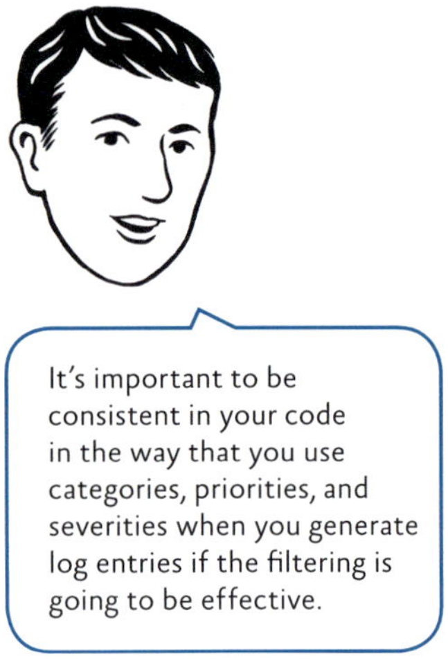

The Semantic Logging Application Block discussed in the next chapter takes a different approach with these properties that makes it much easier to be consistent.

Creating and Using LogEntry Objects

So far we have used the **Write** method of the **LogWriter** class to generate log entries. An alternative approach that may be useful if you want to create log entries individually, perhaps to return them from methods or to pass them between processes, is to generate instances of the **LogEntry** class and then write them to the configured targets afterwards.

The example, *Creating and writing log entries with a LogEntry object*, demonstrates this approach. It creates two **LogEntry** instances. The code first calls the most complex constructor of the **LogEntry** class that accepts all of the possible values. This includes a **Dictionary** of objects with a string key (in this example, the single item **Extra Information**) that will be included in the output of the trace listener and formatter. Then it writes this log entry using an overload of the **Write** method of the **LogWriter** that accepts a **LogEntry** instance.

Next, the code creates a new empty **LogEntry** using the default constructor and populates this by setting individual properties, before writing it using the same **Write** method of the **LogWriter**.

```
// Check if logging is enabled before creating log entries.
if (defaultWriter.IsLoggingEnabled())
{
  // Create a Dictionary of extended properties
  Dictionary<string, object> exProperties = new Dictionary<string, object>();
  exProperties.Add("Extra Information", "Some Special Value");

  // Create a LogEntry using the constructor parameters.
  LogEntry entry1 = new LogEntry("LogEntry with category, priority, event ID, "
                              + "severity, and title.", "General", 8, 9006,
                              TraceEventType.Error, "Logging Block Examples",
                              exProperties);
  defaultWriter.Write(entry1);

  // Create a LogEntry and populate the individual properties.
  LogEntry entry2 = new LogEntry
  {
    Categories = new string[] { "General" },
    EventId = 9007,
    Message = "LogEntry with individual properties specified.",
    Priority = 9,
    Severity = TraceEventType.Warning,
    Title = "Logging Block Examples",
    ExtendedProperties = exProperties
  };
  defaultWriter.Write(entry2);
}
else
{
  Console.WriteLine("Logging is disabled in the configuration.");
}
```

This example writes the log entries to the Windows Application Event Log by using the **General** category. If you view the events this example generates, you will see the values set in the code above including (at the end of the list) the extended property we specified using a **Dictionary**. You can see this in Figure 3.

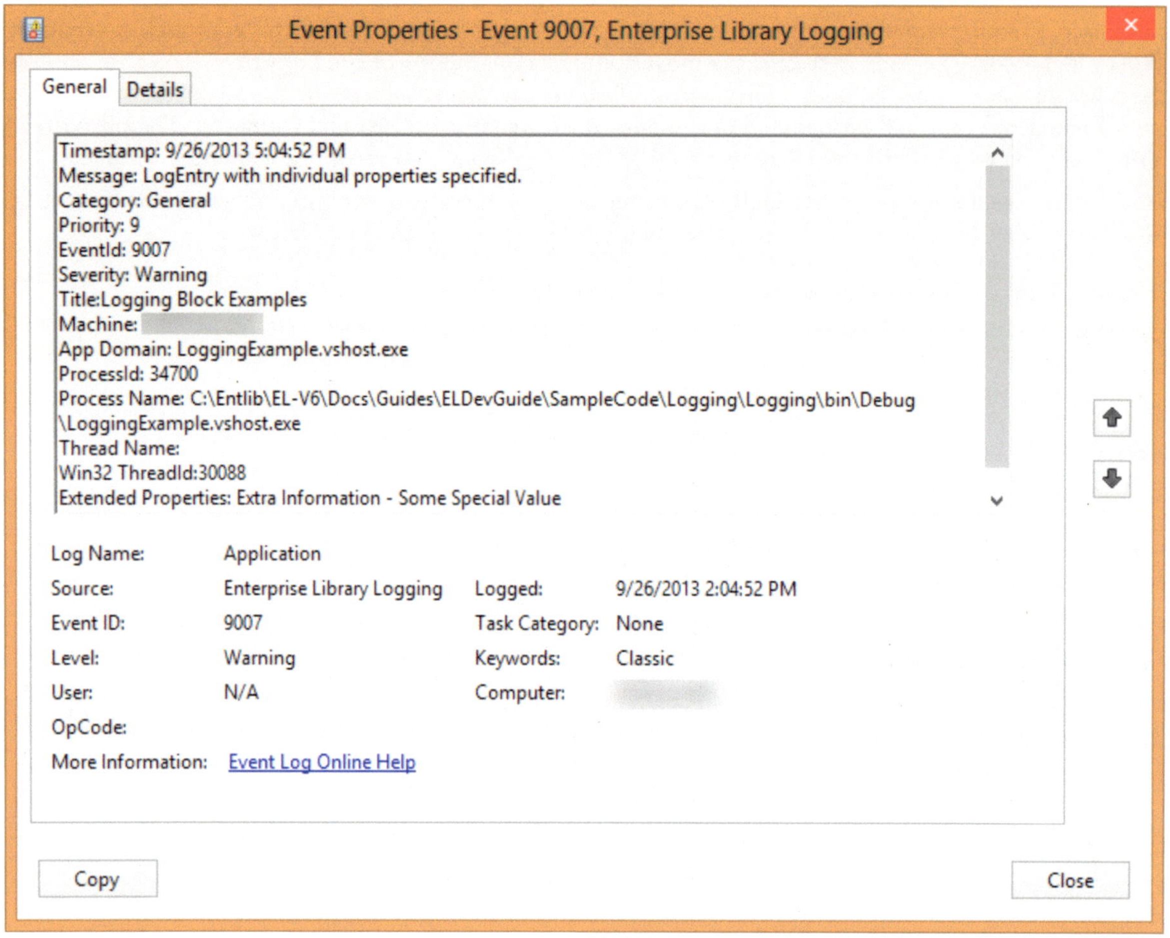

FIGURE 3
A log entry written to the General category

Capturing Unprocessed Events and Logging Errors

The capability to route log entries through different categories to a configured set of trace listener targets provides a very powerful mechanism for performing a wide range of logging activities. However, it prompts some questions. In particular, what happens if the categories specified in a log entry don't match any in the configuration? And what happens if there is an error when the trace listener attempts to write the log entry to the target?

About Special Sources

In fact, the Logging block includes three special sources that handle these situations. Each is effectively a category, and you can add references to configured trace listeners to each one so that events arriving in that category will be written to the target(s) you specify.

The **All Events** special source receives all events, irrespective of all other settings within the configuration of the block. You can use this to provide an audit trail of all events, if required. By default, it has no trace listeners configured.

The **Unprocessed Category** special source receives any log entry that has a category that does not match any configured categories. By default, this category has no trace listeners configured.

When enabled, the **Logging Errors & Warnings** special source receives any log entry that causes an error in the logging process. By default, this category contains a reference to a trace listener that writes details of the error to the Windows Application Event Log, though you can reconfigure this if you wish.

You can use the **Unprocessed Category** as a catch-all mechanism for log entries that weren't handled elsewhere.

An Example of Using Special Sources

The example, *Using Special Sources to capture unprocessed events or errors*, demonstrates how the Logging block reacts under these two circumstances. The code first writes a log entry to a category named **InvalidCategory**, which does not exist in the configuration. Next, it writes another log entry to a category named **CauseLoggingError** that is configured to use a Database trace listener. However, this trace listener specifies a connection string that is invalid; it points to a database that does not exist.

```
// Check if logging is enabled before creating log entries.
if (defaultWriter.IsLoggingEnabled())
{
  // Create log entry to be processed by the "Unprocessed" special source.
  defaultWriter.Write("Entry with category not defined in configuration.",
                      "InvalidCategory");
  // Create log entry to be processed by the "Errors & Warnings" special
  //source.
  defaultWriter.Write("Entry that causes a logging error.", "CauseLoggingError");
}
else
{
  Console.WriteLine("Logging is disabled in the configuration.");
}
```

You might expect that neither of these log entries would actually make it to their target. However, the example generates the following messages that indicate where to look for the log entries that are generated.

```
Created a Log Entry with a category name not defined in the configuration.
The Log Entry will appear in the Unprocessed.log file in the C:\Temp folder.

Created a Log Entry that causes a logging error.
The Log Entry will appear in the Windows Application Event Log.
```

This occurs because we configured the Unprocessed Category with a reference to a flat file trace listener that writes log entries to a file named Unprocessed.log. If you open this file, you will see the log entry that was sent to the **InvalidCategory** category.

The example uses the configuration for the **Logging Errors & Warnings** special source. This means that the log entry that caused a logging error will be sent to the formatted event log trace listener referenced in this category. If you open the application event log, you will see this log entry. The listing below shows some of the content.

```
Timestamp: 24/11/2009 15:14:30
Message: Tracing to LogSource 'CauseLoggingError' failed. Processing for other
sources will continue. See summary information below for more information. Should
this problem persist, stop the service and check the configuration file(s) for
possible error(s) in the configuration of the categories and sinks.

Summary for Enterprise Library Distributor Service:
======================================
-->
Message:
Timestamp: 24/11/2009 15:14:30
Message: Entry that causes a logging error.
Category: CauseLoggingError
...
...
Exception Information Details:
======================================
Exception Type: System.Data.SqlClient.SqlException
Errors: System.Data.SqlClient.SqlErrorCollection
Class: 11
LineNumber: 65536
Number: 4060
Procedure:
Server: (localdb)\v11.0
State: 1
Source: .Net SqlClient Data Provider
ErrorCode: -2146232060
Message: Cannot open database "DoesNotExist" requested by the login. The login failed.
Login failed for user 'xxxxxxx\xxx'.
...
...
StackTrace Information Details:
======================================
...
...
```

In addition to the log entry itself, you can see that the event contains a wealth of information to help you to debug the error. It contains a message indicating that a logging error occurred, followed by the log entry itself. However, after that is a section containing details of the exception raised by the logging mechanism (you can see the error message generated by the **SqlClient** data access code), and after this is the full stack trace.

> *One point to be aware of is that logging database and security exceptions should always be done in such a way as to protect sensitive information that may be contained in the logs. You must ensure that you appropriately restrict access to the logs, and only expose non-sensitive information to other users. You may want to consider applying exception shielding, as described in Chapter 3, "Error Management Made Exceptionally Easy."*

Logging to a Database

One of the most common requirements for logging, after Windows Event Log and text files, is to store log entries in a database. The Logging block contains the database trace listener that makes this easy. You configure the database using a script provided with Enterprise Library, located in the \Blocks\Logging\Src\DatabaseTraceListener\Scripts folder of the source code. We also include these scripts with the example for this chapter.

The scripts assume that you will use the SQL Server Express Local Database instance, but you can edit the **CreateLoggingDb.cmd** file to change the target to a different database server. The SQL script that the command file executes creates a database named **Logging**, and adds the required tables and stored procedures to it.

However, if you only want to run the example application we provide for this chapter, you do not need to create a database. The project contains a preconfigured database file named Logging.mdf (located in the bin\Debug folder) that is auto-attached to your SQL Server LocalDB database instance. You can connect to this database using Visual Studio Server Explorer to see the contents. The constructor of the database trace listener includes a database parameter, which is a reference to this database. The following code sample shows this configuration.

To log to a database you must add the **Enterprise Library 6 - Logging Application Block Database Provider** NuGet package to your project.

```
var databaseTraceListener =
  new FormattedDatabaseTraceListener(
  exampleDatabase, "WriteLog", "AddCategory", extendedFormatter);

config.AddLogSource("Database", SourceLevels.All, true)
     .AddTraceListener(databaseTraceListener);
```

The database trace listener uses a text formatter to format the output, and so you can edit the template used to generate the log message to suit your requirements. You can also add extended properties to the log entry if you wish. In addition, as with all trace listeners, you can filter log entries based on their severity if you like.

The **Log** table in the database contains columns for only the commonly required values, such as the message, event ID, priority, severity, title, timestamp, machine and process details, and more. It also contains a column named **FormattedMessage** that contains the message generated by the text formatter.

The sample application includes the following code to initialize the **DatabaseFactory** static façade from the configuration file.

```
DatabaseProviderFactory factory = new DatabaseProviderFactory(
                              new SystemConfigurationSource(false).GetSection);
DatabaseFactory.SetDatabaseProviderFactory(factory, false);
```

If you want to use declarative configuration you need to bootstrap the **DatabaseFactory** static façade to allow the database listener configuration object to properly set up the listener.

Using the Database Trace Listener

The example, *Sending log entries to a database*, demonstrates the use of the database trace listener. The code is relatively simple, following the same style as the earlier example of creating a **Dictionary** of extended properties, and then using the **Write** method of the **LogWriter** to write two log entries. The first log entry is created by the **LogWriter** from the parameter values provided to the **Write** method. The second is generated in code as a new **LogEntry** instance by specifying the values for the constructor parameters. Also notice how easy it is to add additional information to a log entry using a simple Dictionary as the **ExtendedProperties** of the log entry.

```
// Check if logging is enabled before creating log entries.
if (defaultWriter.IsLoggingEnabled())
{
  // Create a Dictionary of extended properties
  Dictionary<string, object> exProperties = new Dictionary<string, object>();
  exProperties.Add("Extra Information", "Some Special Value");

  // Create a LogEntry using the constructor parameters.
  defaultWriter.Write("Log entry with category, priority, event ID, severity, "
                    + "title, and extended properties.", "Database",
                    5, 9008, TraceEventType.Warning,
                    "Logging Block Examples", exProperties);

  // Create a LogEntry using the constructor parameters.
  LogEntry entry = new LogEntry("LogEntry with category, priority, event ID, "
                             + "severity, title, and extended properties.",
                             "Database", 8, 9009, TraceEventType.Error,
                             "Logging Block Examples", exProperties);
  defaultWriter.Write(entry);
}
else
{
  Console.WriteLine("Logging is disabled in the configuration.");
}
```

To see the two log messages created by this example, you can open the **Logging.mdf** database from the bin\Debug folder using Visual Studio Server Explorer. You will find that the **FormattedMessage** column of the second message contains the following. You can see the extended property information we added using a **Dictionary** at the end of the message.

```
Timestamp: 03/12/2009 17:14:02
Message: LogEntry with category, priority, event ID, severity, title, and extended properties.
Category: Database
Priority: 8
EventId: 9009
Severity: Error
Title: Logging Block Examples
Activity ID: 00000000-0000-0000-0000-000000000000
Machine: BIGFOOT
App Domain: LoggingExample.vshost.exe
ProcessId: 5860
Process Name: E:\Logging\Logging\bin\Debug\LoggingExample.vshost.exe
Thread Name:
Win32 ThreadId:3208
Extended Properties: Extra Information - Some Special Value
```

Note that you cannot simply delete logged information due to the references between the **Log** and **CategoryLog** tables. However, the database contains a stored procedure named **ClearLogs** that you can execute to remove all log entries.

> *The connection string for the database we provide with this example is:*
>
> ```
> Data Source=(localdb)\v11.0;AttachDbFilename=|DataDirectory|\Database\
> Logging.mdf;
> Integrated Security=True
> ```
>
> *If you have configured a different database using the scripts provided with Enterprise Library, you may find that you get an error when you run this example. It is likely to be that you have an invalid connection string in your App.config file for your database. In addition, use the Services applet in your Administrative Tools folder to check that the SQL Server (SQLEXPRESS) database service (the service is named **SQL Server (SQLEXPRESS)**) is running.*

Logging Asynchronously

Where you have high volumes of trace messages or where performance is critical you may want to use the **AsynchronousTraceListenerWrapper** class to wrap an existing listener and enable it to write log messages asynchronously. However, the risk with writing log messages asynchronously is that if the internal buffer overflows or the application shuts down before all the entries are flushed, you will lose some log messages. It is possible to adjust the size of the buffer to your requirements.

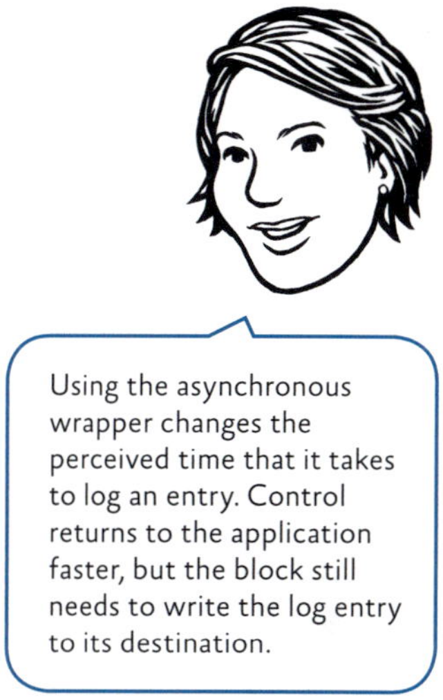

The following code sample shows how you can wrap an existing database listener so that you can write messages asynchronously to the database.

```
var databaseTraceListener = new FormattedDatabaseTraceListener(
    DatabaseFactory.CreateDatabase("ExampleDatabase"),"WriteLog",
                                   "AddCategory",extendedFormatter);

config.AddLogSource("AsyncDatabase", SourceLevels.All, true)
    .AddAsynchronousTraceListener(databaseTraceListener);
```

You can use the asynchronous listener in exactly the same way that you use the wrapped listener. For more information about how to manage the buffer size, see the topic *Logging Messages Asynchronously* in the Enterprise Library Reference Documentation.

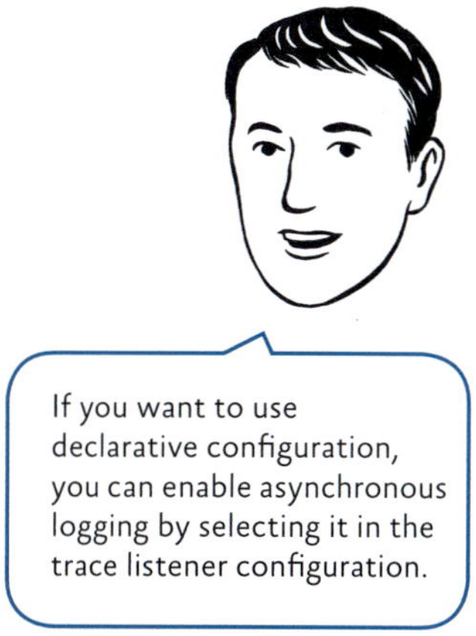

Reconfiguring Logging at Run Time

You assign names to the listeners that you plan to reconfigure, it makes it easier to find them.

If you are troubleshooting a problem with a production system, you may want to be able to make temporary changes to your logging configuration to collect additional information. You may also want to be able to do this without stopping and starting your application.

> *During reconfiguration, all incoming logging operations are blocked, so you want to be as quick as possible.*

The example, *Dynamically changing logging settings*, demonstrates how you can modify the logging configuration settings on the fly. The following code sample shows a method that replaces the filter settings in the configuration. In this example, the only change is to the **maximumPriority** setting in the priority filter.

```
static void ReplacePriorityFilter(int maximumPriority)
{
    defaultWriter.Configure(cfg => {
      cfg.Filters.Clear();
      // Category Filters
      ICollection<string> categories = new List<string>();
      categories.Add("BlockedByFilter");

      // Log Filters
      var priorityFilter = new PriorityFilter(
        "Priority Filter", 2, maximumPriority);
      var logEnabledFilter = new LogEnabledFilter("LogEnabled Filter", true);
      var categoryFilter = new CategoryFilter(
       "Category Filter", categories, CategoryFilterMode.AllowAllExceptDenied);
      cfg.Filters.Add(priorityFilter);
      cfg.Filters.Add(logEnabledFilter);
      cfg.Filters.Add(categoryFilter);
  });
}
```

You can also use the **ConfigurationManager** class to read a configuration value, such as the **maximumPriority** setting from an external source. The **ConfigurationManager** class can monitor the external source for changes and then invoke the **Configure** method. Chapter 1, "Introduction to the Microsoft Enterprise Library," describes how to use the **ConfigurationManager** class.

The **Configure** method in the **LogWriter** class enables you to modify parts of the configuration: in this case the collection of filters in the configuration. You can use the same approach to modify the category sources, special sources, and global configuration settings such as the **IsLoggingEnabled** property.

Testing Logging Filter Status

As you've seen in earlier examples, the Logging block allows you to check if logging is enabled before you create and write a log entry. You can avoid the additional load that this places on your application if logging is not enabled. However, even when logging is enabled, there is no guarantee that a specific log entry will be written to the target log store. For example, it may be blocked by a priority filter if the message priority is below a specified level, or it may belong only to one or more categories where the relevant category filter(s) have logging disabled (a common scenario in the case of logging code specifically designed only for debugging use).

The example, *Checking filter status and adding context information to the log entry,* demonstrates how you can check if a specific log entry will be written to its target before you actually call the **Write** method. After checking that logging is not globally disabled, the example creates two **LogEntry** instances with different categories and priorities. It passes each in turn to another method named **ShowDetailsAndAddExtraInfo**. The following is the code that creates the **LogEntry** instances.

```
// Check if logging is enabled before creating log entries.
if (defaultWriter.IsLoggingEnabled())
{
  // Create a new LogEntry with two categories and priority 3.
  string[] logCategories = new string[] {"General", "DiskFiles"};
  LogEntry entry1 = new LogEntry("LogEntry with categories 'General' and "
                               + "'DiskFiles' and Priority 3.", logCategories,
                               3, 9009, TraceEventType.Error,
                               "Logging Block Examples", null);
  ShowDetailsAndAddExtraInfo(entry1);

  // Create a new LogEntry with one category and priority 1.
  logCategories = new string[] { "BlockedByFilter" };
  LogEntry entry2 = new LogEntry("LogEntry with category 'BlockedByFilter' and "
                               + "Priority 1.", logCategories, 1, 9010,
                               TraceEventType.Information,
                               "Logging Block Examples", null);
  ShowDetailsAndAddExtraInfo(entry2);
}
else
{
  Console.WriteLine("Logging is disabled in the configuration.");
}
```

The **ShowDetailsAndAddExtraInfo** method takes a **LogEntry** instance and does two different things. Firstly, it shows how you can obtain information about the way that the Logging block will handle the log entry. This may be useful in advanced scenarios where you need to be able to programmatically determine if a specific log entry was detected by a specific trace source, or will be written to a specific target. Secondly, it demonstrates how you can check if specific filters, or all filters, will block a log entry from being written to its target.

Obtaining Information about Trace Sources and Trace Listeners

The first section of the **ShowDetailsAndAddExtraInfo** method iterates through the collection of trace sources (**LogSource** instances) exposed by the **GetMatchingTraceSources** method of the **LogWriter** class. Each **LogSource** instance exposes a **Listeners** collection that contains information about the listeners (which specify the targets to which the log entry will be sent).

```
void ShowDetailsAndAddExtraInfo(LogEntry entry)
{
  // Display information about the Trace Sources and Listeners for this //
  // LogEntry.
  IEnumerable<LogSource> sources = defaultWriter.
                                   GetMatchingTraceSources(entry);
  foreach (LogSource source in sources)
  {
    Console.WriteLine("Log Source name: '{0}'", source.Name);
    foreach (TraceListener listener in source.Listeners)
    {
      Console.WriteLine(" - Listener name: '{0}'", listener.Name);
    }
  }
  ...
```

Checking if Filters Will Block a Log Entry

To use this feature, you need to know which filters exist before you can perform the check.

Next, the **ShowDetailsAndAddExtraInfo** method checks if any filters will block the current log entry. There are two ways you can do this. You can query each filter type in turn, or just a specific filter type, by using the **GetFilter** method of the **LogWriter** class to get a reference to that type of filter. Then you can check if this filter is enabled, and also use the **ShouldLog** method (to which you pass the list of categories for the log entry) to see if logging will succeed.

The following code shows this approach. It also shows the simpler approach that you can use if you are not interested in the result for a specific filter type. The **LogWriter** class also exposes the **ShouldLog** method, which indicates if any filters will block this entry.

```
...
  // Check if any filters will block this LogEntry.
  // This approach allows you to check for specific types of filter.
  // If there are no filters of the specified type configured, the GetFilter
  // method returns null, so check this before calling the ShouldLog method.
  CategoryFilter catFilter = defaultWriter.GetFilter<CategoryFilter>();

  if (null == catFilter || catFilter.ShouldLog(entry.Categories))
  {
    Console.WriteLine("Category Filter(s) will not block this LogEntry.");
  }
  else
  {
    Console.WriteLine("A Category Filter will block this LogEntry.");
  }

  PriorityFilter priFilter = defaultWriter.GetFilter<PriorityFilter>();
  if (null == priFilter || priFilter.ShouldLog(entry.Priority))
  {
    Console.WriteLine("Priority Filter(s) will not block this LogEntry.");
  }
```

```
    else
    {
      Console.WriteLine("A Priority Filter will block this LogEntry.");
    }

    // Alternatively, a simple approach can be used to check for any type of filter
    if (defaultWriter.ShouldLog(entry))
    {
      Console.WriteLine("This LogEntry will not be blocked by config settings.");
      ...
      // Add context information to log entries after checking that the log entry
      // will not be blocked due to configuration settings. See the following
      // section 'Adding Additional Context Information' for details.
      ...
    }
    else
    {
      Console.WriteLine(
                    "This LogEntry will be blocked by configuration settings.");
    }
  }
```

After you determine that logging will succeed, you can add extra context information and write the log entry. You'll see the code to achieve this shortly. In the meantime, this is the output generated by the example. You can see that it contains details of the log (trace) sources and listeners for each of the two log entries created by the earlier code, and the result of checking if any category filters will block each log entry.

By checking whether the logging will succeed, you can decide whether to perform additional operations such as collecting context information to add to the log entry which might be relatively expensive.

```
Created a LogEntry with categories 'General' and 'DiskFiles'.
Log Source name: 'General'
 - Listener name: 'Formatted EventLog TraceListener'
Log Source name: 'DiskFiles'
 - Listener name: 'FlatFile TraceListener'
 - Listener name: 'XML Trace Listener'
Category Filter(s) will not block this LogEntry.
Priority Filter(s) will not block this LogEntry.
This LogEntry will not be blocked due to configuration settings.
...
Created a LogEntry with category 'BlockedByFilter', and Priority 1.
Log Source name: 'BlockedByFilter'
 - Listener name: 'Formatted EventLog TraceListener'
A Category Filter will block this LogEntry.
A Priority Filter will block this LogEntry.
This LogEntry will be blocked due to configuration settings.
```

Adding Additional Context Information

While it's useful to have every conceivable item of information included in your log messages, it's not always the best approach. Collecting information from the environment absorbs processing cycles and increases the load that logging places on your application. The Logging block is highly optimized to minimize the load that logging incurs. As an example, some of the less useful information is not included in the log messages by default—particularly information that does require additional resources to collect.

However, you can collect this information if you wish. You may decide to do so in special debugging instrumentation that you only turn on when investigating problems, or for specific areas of your code where you need the additional information, such as security context details for a particularly sensitive process.

After checking that a log entry will not be blocked by filters, the **ShowDetailsAndAddExtraInfo** method (shown in the previous section) adds a range of additional context and custom information to the log entry. It uses the four standard Logging block helper classes that can generate additional context information and add it to a **Dictionary**. These helper classes are:

- The **DebugInformationProvider**, which adds the current stack trace to the **Dictionary**.
- The **ManagedSecurityContextInformationProvider**, which adds the current identity name, authorization type, and authorization status to the **Dictionary**.
- The **UnmanagedSecurityContextInformationProvider**, which adds the current user name and process account name to the **Dictionary**.
- The **ComPlusInformationProvider**, which adds the current activity ID, application ID, transaction ID (if any), direct caller account name, and original caller account name to the **Dictionary**.

The following code shows how you can use these helper classes to create additional information for a log entry. It also demonstrates how you can add custom information to the log entry—in this case by reading the contents of the application configuration file into the **Dictionary**. After populating the **Dictionary**, you simply set it as the value of the **ExtendedProperties** property of the log entry before writing that log entry.

```
...
// Create additional context information to add to the LogEntry.
Dictionary<string, object> dict = new Dictionary<string, object>();
// Use the information helper classes to get information about
// the environment and add it to the dictionary.
DebugInformationProvider debugHelper = new DebugInformationProvider();
debugHelper.PopulateDictionary(dict);

ManagedSecurityContextInformationProvider infoHelper
                                    = new ManagedSecurityContextInformationProvider();
infoHelper.PopulateDictionary(dict);

UnmanagedSecurityContextInformationProvider secHelper
                                    = new UnmanagedSecurityContextInformationProvider();
secHelper.PopulateDictionary(dict);

ComPlusInformationProvider comHelper = new ComPlusInformationProvider();
comHelper.PopulateDictionary(dict);

// Get any other information you require and add it to the dictionary.
string configInfo = File.ReadAllText(@"..\..\App.config");
dict.Add("Config information", configInfo);

// Set dictionary in the LogEntry and write it using the default LogWriter.
entry.ExtendedProperties = dict;
defaultWriter.Write(entry);
...
```

The example produces the following output on screen.

```
Added the current stack trace to the Log Entry.
Added current identity name, authentication type, and status to the Log
Entry.
Added the current user name and process account name to the Log Entry.
Added COM+ IDs and caller account information to the Log Entry.
Added information about the configuration of the application to the Log
Entry.
LogEntry written to configured trace listeners.
```

To see the additional information added to the log entry, open Windows Event Viewer and locate the new log entry. We haven't shown the contents of this log entry here as it runs to more than 350 lines and contains just about all of the information about an event occurring in your application that you could possibly require!

Tracing and Correlating Activities

The final topic for this chapter demonstrates another feature of the Logging block that makes it easier to correlate multiple log entries when you are trying to trace or debug some recalcitrant code in your application. One of the problems with logging is that relying simply on the event ID to correlate multiple events that are related to a specific process or section of code is difficult and error prone. Event IDs are often not unique, and there can be many events with the same event ID generated from different instances of the components in your application that are intermixed in the logs: using the Semantic Logging Application Block (described in the next chapter) helps you to address this specific issue.

The Logging block makes it easy to add an additional unique identifier to specific log entries that you can later use to filter the log and locate only entries related to a specific process or task. The Logging block tracing feature makes use of the .NET Correlation Manager class, which maintains an Activity ID that is a GUID. By default, this is not set, but the Logging block allows you to use a **TraceManager** to generate Tracer instances. Each of these sets the Activity ID to a user supplied or randomly generated GUID value that is maintained only during the context of the tracer. The Activity ID returns to its previous value when the tracer is disposed or goes out of scope.

You specify an operation name when you create the tracer. This is effectively the name of a category defined in the configuration of the block. All log entries created within the context of the tracer will be assigned to that category in addition to any categories you specify when you create the log entry.

You can specify a GUID value when you create and start a tracer, and all subsequent log entries within the scope of that tracer and all nested tracers that do not specify a different GUID will have the specified activity ID. If you start a new nested tracer instance within the scope of a previous one, it will have the same activity ID as the parent tracer unless you specify a different one when you create and start the nested tracer; in that case, this new activity ID will be used in subsequent log entries within the scope of this tracer. Tracing and activity IDs rely on thread-affinity which won't be guaranteed in asynchronous scenarios such as when you use **Tasks**.

Often, you want to correlate the events for a specific instance of an operation. The more independent each instance is, the more valuable it is to narrow your view of events to those that are related.

Although the Logging block automatically adds the activity ID to each log entry, this does not appear in the resulting message when you use the text formatter with the default template. To include the activity ID in the logged message that uses a text formatter, you must edit the template property in the configuration tools to include the token {**property(ActivityId)**}. Note that property names are case-sensitive in the template definition.

An Example of Tracing Activities

The example, *Tracing activities and publishing activity information to categories*, should help to make this clear. At the start of the application, the code creates a **TraceManager** instance.

```
LoggingConfiguration loggingConfiguration = BuildProgrammaticConfig();
LogWriter defaultWriter = new LogWriter(loggingConfiguration);

// Create a TraceManager object.
TraceManager traceMgr = new TraceManager(defaultWriter);
```

Next, the code creates and starts a new **Tracer** instance using the **StartTrace** method of the **TraceManager**, specifying the category named **General**. As it does not specify an Activity ID value, the **TraceManager** creates one automatically (assuming it doesn't have one already). This is the preferred approach, because each separate process running an instance of this code will generate a different GUID value. This means you can isolate individual events for each process.

The code then creates and writes a log entry within the context of this tracer, specifying that it belongs to the **DiskFiles** category in addition to the **General** category defined by the tracer. Next, it creates a nested **Tracer** instance that specifies the category named **Database**, and writes another log entry that itself specifies the category named **Important**. This log entry will therefore belong to the **General**, **Database**, and **Important** categories. Then, after the **Database** tracer goes out of scope, the code creates a new **Tracer** that again specifies the **Database** category, but this time it also specifies the Activity ID to use in the context of this new tracer. Finally, it writes another log entry within the context of the new **Database** tracer scope.

```
// Start tracing for category 'General'. All log entries within trace context
// will be included in this category and use any specified Activity ID (GUID).
// If you do not specify an Activity ID, the TraceManager will create a new one.
using (traceMgr.StartTrace("General"))
{
  // Write a log entry with another category, will be assigned to both.
  defaultWriter.Write("LogEntry with category 'DiskFiles' created within "
                      + "context of 'General' category tracer.", "DiskFiles");

  // Start tracing for category 'Database' within context of 'General' tracer.
  // Do not specify a GUID to use so that the existing one is used.
  using (traceMgr.StartTrace("Database"))
  {
    // Write a log entry with another category, will be assigned to all three.
    defaultWriter.Write("LogEntry with category 'Important' created within "
                        + "context of first nested 'Database' category tracer.", "Important");
  }
```

```
    // Back in context of 'General' tracer here.
    // Start tracing for category 'Database' within context of 'General' tracer
    // as above, but this time specify a GUID to use.
    using (traceMgr.StartTrace("Database", new Guid("{12345678-1234-1234-1234-123456789ABC}")))

    {
      // Write a log entry with another category, will be assigned to all three.
      defaultWriter.Write("LogEntry with category 'Important' created within "
                          + "context of nested 'Database' category tracer.", "Important");
    }
    // Back in context of 'General' tracer here and back to the
    // randomly created activity ID.
  }
```

Not shown above are the lines of code that, at each stage, write the current Activity ID to the screen. The output generated by the example is shown here. You can see that, initially, there is no Activity ID. The first tracer instance then sets the Activity ID to a random value (you will get a different value if you run the example yourself), which is also applied to the nested tracer.

However, the second tracer for the **Database** category changes the Activity ID to the value we specified in the **StartTrace** method. When this tracer goes out of scope, the Activity ID is reset to that for the parent tracer. When all tracers go out of scope, the Activity ID is reset to the original (empty) value.

```
 - Current Activity ID is: 00000000-0000-0000-0000-000000000000

Written LogEntry with category 'DiskFiles' created within context of 'General'
 category tracer.

 - Current Activity ID is: a246ada3-e4d5-404a-bc28-4146a190731d

Written LogEntry with category 'Important' created within context of first
 'Database' category tracer nested within 'DiskFiles' category TraceManager.

 - Current Activity ID is: a246ada3-e4d5-404a-bc28-4146a190731d

Leaving the context of the first Database tracer

 - Current Activity ID is: a246ada3-e4d5-404a-bc28-4146a190731d

Written LogEntry with category 'Important' created within context of second
 'Database' category tracer nested within 'DiskFiles' category TraceManager.

 - Current Activity ID is: 12345678-1234-1234-1234-123456789abc

Leaving the context of the second Database tracer

 - Current Activity ID is: a246ada3-e4d5-404a-bc28-4146a190731d

Leaving the context of the General tracer

 - Current Activity ID is: 00000000-0000-0000-0000-000000000000

Open the log files in the folder C:\Temp to see the results.
```

If you open the RollingFlatFile.log file you will see the two log entries generated within the context of the nested tracers. These belong to the categories Important, Database, and General. You will also see the Activity ID for each one, and can confirm that it is different for these two entries. For example, this is the first part of the log message for the second nested tracer, which specifies the Activity ID GUID in the **StartTrace** method.

```
Timestamp: 01/12/2009 12:12:00
Message: LogEntry with category 'Important' created within context of second nested 'Database' category
tracer.
Category: Important, Database, General
Priority: -1
EventId: 1
Severity: Information
Title:
Activity ID: 12345678-1234-1234-1234-123456789abc
```

Be aware that other software and services may use the Activity ID of the Correlation Manager to provide information and monitoring facilities. An example is Windows Communication Foundation (WCF), which uses the Activity ID to implement tracing.

You must also ensure that you correctly dispose ***Tracer*** *instances. If you do not take advantage of the* ***using*** *construct to automatically dispose instances, you* must *ensure that you dispose nested instances in the reverse order you created them—by disposing the child instance before you dispose the parent instance. You must also ensure that you dispose Tracer instances on the same thread that created them.*

You must be careful in asynchronous scenarios such as when you use the ***await*** *keyword: this makes it easy to forget that the disposal of a tracer created in a* ***using*** *statement could happen in another thread.*

Creating Custom Trace Listeners, Filters, and Formatters

You can extend the capabilities of the Logging block if you need to add specific functionality to it. In general, you will only need to implement custom log filters, trace listeners, or log formatters. The design of the block makes it easy to add these and make them available through configuration.

To create a new log filter, you can either implement the **ILogFilter** interface, which specifies the single method **Filter** that must accept an instance of a **LogEntry** and return true or false, or you can inherit the base class **LogFilter** and implement the **Filter** method.

To create a custom trace listener, you can inherit from the abstract base class **CustomTraceListener** and implement the methods required to send your log entry to the appropriate location or execute the relevant actions to log the message. You can expose a property for the relevant log formatter if you want to allow users to select a specific formatter for the message.

To create a custom log formatter, you can either implement the ILogFormatter interface, which specifies the single method, **Format**, that must accept an instance of a **LogEntry** and return the formatted message, or you can inherit the base class, **LogFormatter**, and implement the **Format** method.

For more information about using and extending the Logging block, see *The Logging Application Block* in the Enterprise Library Reference Documentation.

Summary

This chapter described the Enterprise Library Logging Application Block. This block is extremely useful for logging activities, events, messages, and other information that your application must persist or expose—both to monitor performance and to generate auditing information. The Logging block is, like all of the other Enterprise Library blocks, highly customizable and driven through either programmatic or declarative configuration so that you (or administrators and operations staff) can modify the behavior to suit your requirements exactly.

You can use the Logging block to categorize, filter, and write logging information to a wide variety of targets, including Windows event logs, email messages, disk files, Windows Message Queuing, and a database. You can even collect additional context information and add it to the log entries automatically, and add activity IDs to help you correlate related messages and activities. And, if none of the built-in features meets your requirements, you can create and integrate custom listeners, filters, and formatters.

This chapter explained why you should consider decoupling your logging features from your application code, what the Logging block can do to help you implement flexible and configurable logging, and how you actually perform the common tasks related to logging. For more information about using the Logging block, see the *Enterprise Library Reference Documentation*.

Before deciding to use the Logging Application Block in your application, you should consider using the Semantic Logging Application Block. The Semantic Logging Application Block offers a more structured approach to logging and when used out-of-process further reduces the overhead of logging on your application.

More Information

All links in this book are accessible from the book's online bibliography on MSDN at *http://aka.ms/el6biblio*.

For more information about using and extending the Logging block, see *The Logging Application Block* in the Enterprise Library Reference Documentation.

For more information about how to manage the buffer size, see the topic *Logging Messages Asynchronously* in the Enterprise Library Reference Documentation.

General Links:

- There are resources to help if you're getting started with Enterprise Library, and there's help for existing users as well (such as the breaking changes and migration information for previous versions) available on the Enterprise Library Community Site at *http://www.codeplex.com/entlib/*.
- For more details about the Enterprise Library Application Blocks, see the *Enterprise Library Reference Documentation* and the *Enterprise Library 6 Class Library*.
- You can download and work through the Hands-On Labs for Enterprise Library, which are available at *http://aka.ms/el6hols*.
- You can download the Enterprise Library binaries, source code, Reference Implementation, or QuickStarts from the Microsoft Download Center at *http://go.microsoft.com/fwlink/p/?LinkID=290898*.
- To help you understand how you can use Enterprise Library application blocks, we provide a series of simple example applications that you can run and examine. To obtain the example applications, go to *http://go.microsoft.com/fwlink/p/?LinkID=304210*.

6 Logging What You Mean
Using the Semantic Logging Application Block

Introduction

Why do you need another logging block when the Logging Application Block already has a comprehensive set of features? You'll find that you can do many of the same things with the Semantic Logging Application Block: you can write log messages to multiple destinations, you can control the format of your log messages, and you can filter what gets written to the log. You also use the Semantic Logging Application Block for the same reasons: collecting diagnostic information from your application, debugging, and troubleshooting. The answer why you might want to use the Semantic Logging Application Block lies with how the application writes messages to the logs.

Using a traditional logging infrastructure, such as that provided by the Logging Application Block, you might record an event entry of your application as shown in the following code sample:

```
LogEntry logEntry = new LogEntry();
logEntry.EventId = 100;
logEntry.Severity = TraceEventType.Information;
logEntry.Message = String.Format(
          @"Scaling request for role {0} has being successfully submitted.
          The requested role instance count is {1}.
          The scaling operation was triggered by a rule named '{2}'.
          The current role instance count is {3}.",
          request.RoleName,
          request.InstanceCount,
          context.RuleName,
          context.CurrentInstanceCount);

logEntry.Categories.Add("Autoscaling Updates");

logWriter.Write(logEntry);
```

Using the Semantic Logging Application Block, you would record the same entry as shown in the following code sample:

```
MyCompanyEventSource.Log.ScalingRequestSubmitted(
                         request.RoleName,
                         request.InstanceCount,
                         context.RuleName,
                         context.CurrentInstanceCount);
```

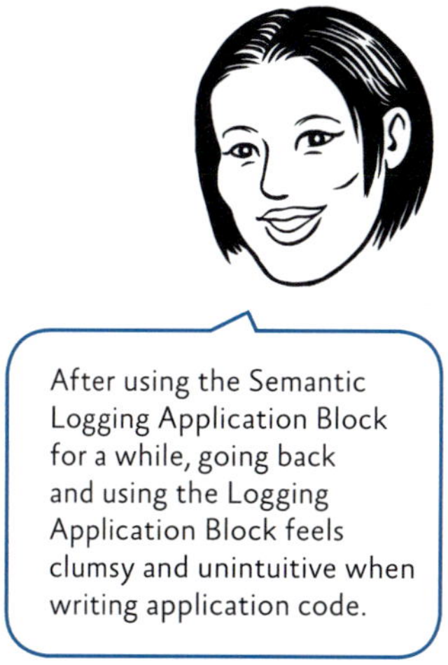

Notice how with the Semantic Logging Application Block you are simply reporting the fact that some event occurred that you might want to record in a log. You do not need to specify an event ID, a severity, or the format of the message when you create the event to log in your application code. This approach of using strongly typed events in your logging process provides the following benefits:

- You can be sure that you format and structure your log messages in a consistent way because there is no chance of making a coding error when you write the log message. For example, an event with a particular ID will always have the same verbosity, extra information, and payload structure.
- It is easier to have a holistic view of all the application events, as they are defined in a centralized place, and refactor to make them consistent with each other.
- It is easier to query and analyze your log files because the log messages are formatted and structured in a consistent manner.
- You can more easily parse your log data using some kind of automation. This is especially true if you are using one of the Semantic Logging Application Block sinks that preserves the structure of the payload of the events; for example, the Windows Azure Table sink preserves the structure of your log messages. However, even if you are using flat files, you can be sure that the format of log messages with a particular ID will be consistent.
- You can more easily consume the log data from another application. For example, in an application that automates activities in response to events that are recorded in a log file or a Windows Azure Table.
- It is easier to correlate log entries from multiple sources.

> *The term **semantic logging** refers specifically to the use of strongly typed events and the consistent structure of the log messages in the Semantic Logging Application Block.*

While it is possible to use the **EventSource** class in the .NET framework (without the Semantic Logging Application Block) and *Event Tracing for Windows* (ETW) to write log messages from your application, using the Semantic Logging Application Block makes it easier to incorporate the functionality provided by the **EventSource** class, and makes it easier to manage the logging behavior of your application. The Semantic Logging Application Block makes it possible to write log messages to multiple destinations, such as flat files or a database, and makes it easier to access the capabilities of the **EventSource** class and control what your application logs by setting filters and logging verbosity. If you are using the block in-process, it does not use the ETW infrastructure to handle log messages but it does have the same semantic approach to logging as ETW. If you are using the block out-of-process, then it uses the ETW infrastructure in Windows to pass the log messages from your application to the sinks that process and save the log messages.

The Semantic Logging Application Block is intended to help you move from the traditional, predominantly unstructured or quasi-structured logging approach (such as that offered by the Logging Application Block) towards the semantic logging approach offered by ETW. With the in-process approach, you don't have to buy in to the full ETW infrastructure but it does enable you to start using semantic logging. The Semantic Logging Application Block enables you to use the **EventSource** class and semantic log messages in your applications without moving away from the log formats you are familiar with (such as flat files and database tables). In the future, you can easily migrate to a complete ETW-based solution without modifying your application code: you continue to use the same custom Event Source class, but use ETW tooling to capture and process your log messages instead of using the Semantic Logging Application Block event listeners.

You can think of the Semantic Logging Application Block as a stepping stone from a traditional logging approach (such as that taken by the Logging Application Block), to a modern, semantic approach as provided by ETW. Even if you never migrate to using the ETW infrastructure, there's plenty of value in adopting a semantic approach to your logging.

You can also use the Semantic Logging Application Block to create an out-of-process logger, an approach that utilizes the ETW infrastructure in Windows. The key advantages of the out-of-process approach are that it makes logging more resilient to application crashes and facilitates correlating log data from multiple sources. If your application crashes, any log messages are likely to be already buffered in the ETW infrastructure and so will be processed and saved. This contrasts with the in-process approach where log messages are buffered and processed within your own application.

Many of the features of the Semantic Logging Application Block are similar to those of the Logging Application Block, and this chapter will refer to Chapter 5, "As Easy As Falling Off a Log," where appropriate.

Collecting trace messages from your production system in a separate, out-of-process, application helps to improve the resiliency of your logging processes.

What Does the Semantic Logging Application Block Do?

When used in-process, the Semantic Logging Application Block enables you to use the **EventSource** class to write log messages from your application. The Semantic Logging Application Block receives notifications whenever the application writes a message using an **EventSource** class. The Semantic Logging Application Block then writes the message to one or more destinations of your choice. The Semantic Logging Application Block includes event sinks that can send log messages to a flat file, a console window, a database, or Windows Azure storage.

Figure 1 illustrates how your application uses the Semantic Logging Application Block in-process. It uses a custom class that extends the **EventSource** class in the **System.Diagnostics.Tracing** namespace to enable you to write application-specific log messages. The event source then notifies the event listener in your application when there is a log message to handle, and then the event sinks, that you attach to the listener, save the log message. The event sinks write the log message to a destination such as file, a database table, or a Windows Azure storage table. All of this takes place in process using managed code.

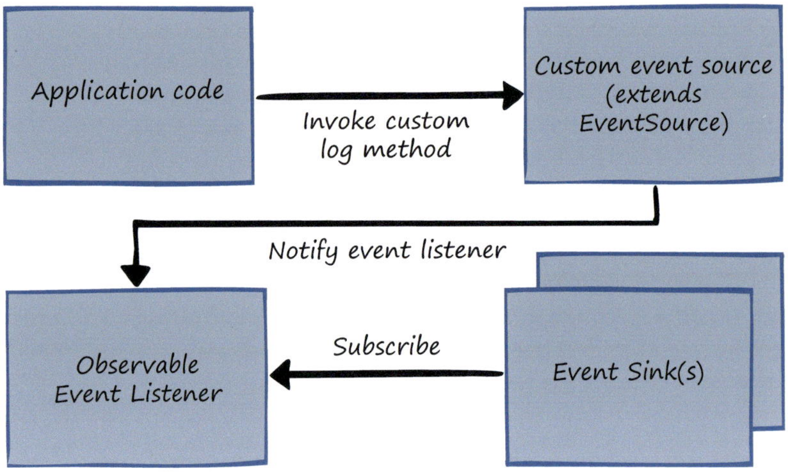

FIGURE 1
The Semantic Logging Application Block In-Process Scenario

Typically, you have a single custom event source class, but you can have more if needed. Depending on how you choose to attach the sinks to the listeners, it's possible that you will have one event listener per sink.

Figure 2 illustrates how you can use the Semantic Logging Application Block out-of-process. This is more complex to configure, but has the advantage of improving the resiliency of logging in your LOB application. This approach uses managed code in the process that sends the log messages and in the process that collects and processes the log messages; some unmanaged code from the operating system is responsible for delivering the log messages between the processes.

In the out-of-process scenario, both the LOB application that generates log messages and the logging application that collects the messages must run on the same machine.

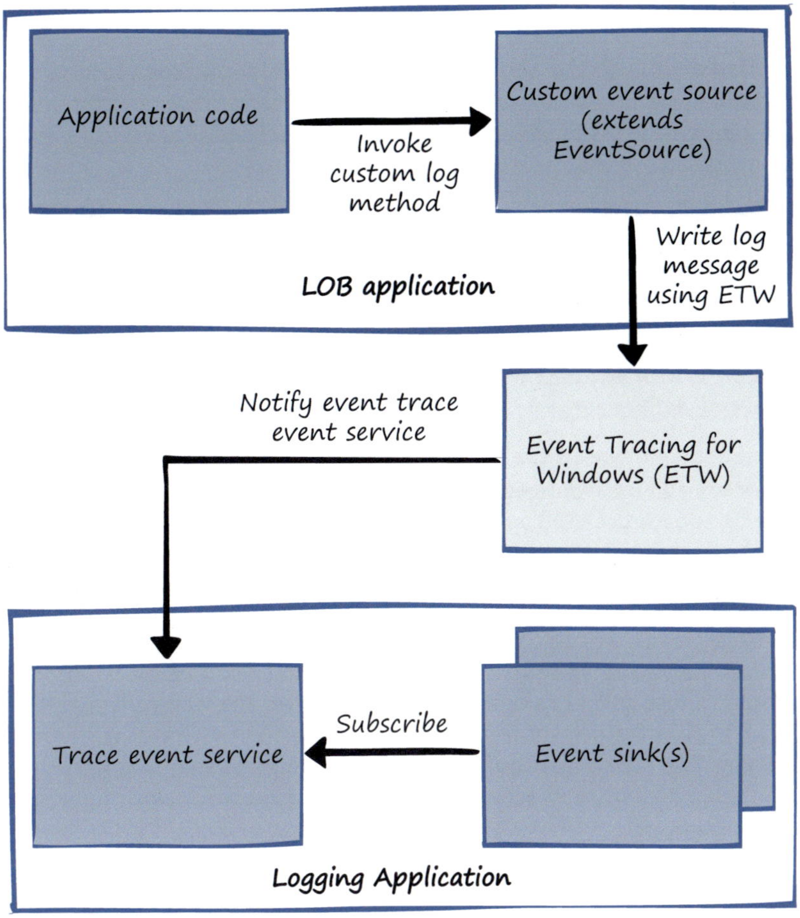

Figure 2
Using Semantic Logging Application Block out-of-process

The Semantic Logging Application Block can work in-process or out-of-process: you can host the event listeners and sinks in your LOB application process or in a separate logging process. Either way you can use the same event sink and formatter types.

In-Process or Out-of-Process?

When should you use the Semantic Logging Application Block in-process and when should you use it out-of-process? Although it is easier to configure and use the block in-process, there are a number of advantages to using it in the out-of-process scenario.

You could still lose log messages in the out-of-process scenario if the server itself fails between the time the application writes the log message and the time the out-of-process host persists the message, or if the out-of-process host application crashes before it persists the message. However, out-of-process host is a robust and relatively simple application and is unlikely to crash.

Using the block out-of-process minimizes the risk of losing log messages if your line-of-business application crashes. When the line-of-business application creates a log message it immediately delivers it to the ETW infrastructure in the operating system, so that any log messages written by the line-of-business application will not get lost if that application crashes.

> *Notice, with very high throughput, you may lose messages if either the ETW or sink buffers become full.*

One further advantage of the out-of-process approach is that it makes it much easier for an administrator to change the configuration at run time without needing to update the LOB application.

Buffering Log Messages

Some sinks, such as the **WindowsAzureTableSink** and **SqlDatabaseSink** classes can buffer log messages for a configurable period of time (ten seconds by default). These sinks typically communicate over the network with the service that is ultimately responsible for persisting the log messages from your application. The block uses buffering in these sinks to improve performance: typically, chunky rather chatty communication over a network improves the overall throughput.

However, using buffering introduces a trade-off: if the process that is buffering the messages crashes before delivering those log messages, then you lose those messages. The shorter the buffering period, the fewer messages you will lose in the event of an application crash, but at the cost of a lower throughput of log messages.

One of the reasons for using the Semantic Logging Application Block out-of-process is to mitigate this risk. If the message buffer is in a separate process from your line-of-business application, then the messages are not lost in the event that the application crashes. The out-of-process host applications for Semantic Logging Application Block sinks are designed to be robust in order to minimize the chances of these applications crashing.

If you try to push a very high volume of events (in the order of magnitude of 100s or 1000s per second using the Windows Azure sink), you may lose some log messages.

You should monitor the log messages generated by the Semantic Logging Application Block for any indication that the have buffers overflowed and that you have lost messages. For example, log messages with event ids 900 and 901 indicate that a sink's internal buffers have overflowed; in the out-of-process scenario, event ids 806 and 807 indicate that the ETW buffers have overflowed. You can modify the buffering configuration options for the sinks to reduce the chance that the buffers overflow with your typical workloads.

For more information, see the topic *Performance Considerations* in the Enterprise Library Reference Documentation.

How Do I Use the Semantic Logging Application Block?

It's time to see some examples of the Semantic Logging Application Block in use, including how to create an event source, how to configure the block, and how to write log entries. The code samples included in this section are taken from the sample application (Semantic Logging) that accompanies this chapter: in some cases, the code is shown here differs slightly from the code in the sample to make it easier to read.

Creating an Event Source

Before you can write log messages using an **EventSource** class, you must define what log messages you will write. This is what semantic logging means. You can specify the log messages you will write by extending the **EventSource** class in the **System.Diagnostics.Tracing** namespace in the .NET 4.5 framework. In the terms used by ETW, an **EventSource** implementation represents an "Event Provider."

Each event type in your application is represented by a method in your **EventSource** implementation. These event methods take parameters that define the payload of the event and are decorated with attributes that provide additional metadata such as the event ID or verbosity level.

The following code sample shows an example **EventSource** implementation. This example also includes the optional nested classes **Keywords** and **Tasks** that enable you to define additional information for your log messages.

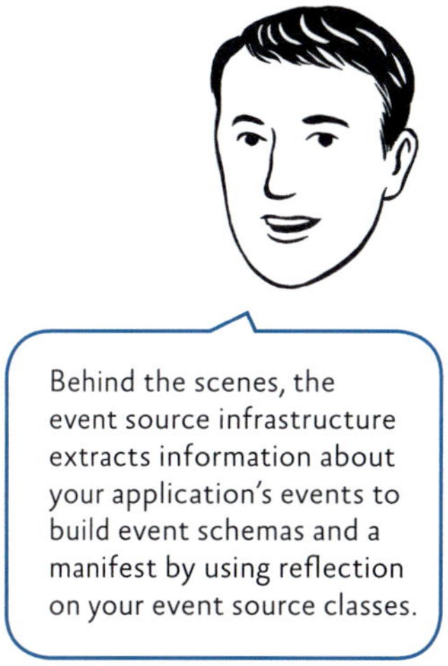


```
[EventSource(Name = "MyCompany")]
public class MyCompanyEventSource : EventSource
{
    public class Keywords
    {
        public const EventKeywords Page = (EventKeywords)1;
        public const EventKeywords DataBase = (EventKeywords)2;
        public const EventKeywords Diagnostic = (EventKeywords)4;
        public const EventKeywords Perf = (EventKeywords)8;
    }

    public class Tasks
    {
        public const EventTask Page = (EventTask)1;
        public const EventTask DBQuery = (EventTask)2;
    }

    private static MyCompanyEventSource _log = new MyCompanyEventSource();
    private MyCompanyEventSource() { }
    public static MyCompanyEventSource Log { get { return _log; } }

    [Event(1, Message = "Application Failure: {0}",
    Level = EventLevel.Critical, Keywords = Keywords.Diagnostic)]
    internal void Failure(string message)
    {
      this.WriteEvent(1, message);
    }
```

```
    [Event(2, Message = "Starting up.", Keywords = Keywords.Perf,
           Level = EventLevel.Informational)]
    internal void Startup()
    {
      this.WriteEvent(2);
    }

    [Event(3, Message = "loading page {1} activityID={0}",
           Opcode = EventOpcode.Start,
           Task = Tasks.Page, Keywords = Keywords.Page,
           Level = EventLevel.Informational)]
    internal void PageStart(int ID, string url)
    {
      if (this.IsEnabled()) this.WriteEvent(3, ID, url);
    }
    ...
}
```

The class, **MyCompanyEventSource** that extends the **EventSource** class, contains definitions for all of the events that you want to be able to log using ETW; each event is defined by its own method. You can use the **EventSource** attribute to provide a more user friendly name for this event source that ETW will use when you save log messages.

Each event is defined using a method that wraps a call to the **WriteEvent** method in the **EventSource** class. The first parameter of the **WriteEvent** method is an event id that must be unique to that event, and different overloaded versions of this method enable you to write additional information to the log. Attributes on each event method further define the characteristics of the event. Notice that the **Event** attribute includes the same value as its first parameter.

In the **MyCompanyEventSource** class, the methods such as **PageStart** include a call to the **IsEnabled** method in the parent class to determine whether to write a log message. The **IsEnabled** method helps to improve performance if you are using the costly overload of the **WriteMethod** that includes a **params object[]** parameter.

There is also an overload of the **IsEnabled** method that can check whether the event source is enabled for particular keywords and levels. However, with this overload, it's important to make sure that the parameters to the **IsEnabled** method match the keywords and levels in the **EventAttribute** attribute. Checking that the parameters match is one of the checks that the **EventSourceAnalyzer** performs: this class is described later in this chapter.

You can use the **Event** attribute to further refine the characteristics of specific events.

For more information about the **EventSource** class and the **WriteEvent** and **IsEnabled** methods, see *EventSource Class* on MSDN.

This **EventSource** follows the recommended naming convention. The name should start with your company name, and if you expect to have more than one **event source** for your company the name should include categories separated by '-'. Microsoft follows this convention; for example, there is an event source called "Microsoft-Windows-DotNetRuntime." You should carefully consider this name because if you change it in the future, users consuming the events will no longer receive them.

The custom event source file can get quite large; you should consider using partial classes to make it more manageable.

You can also include the call to the **IsEnabled** method in a non-event method in your event source class that performs some pre-processing, transforming, or formatting of data before calling an event method.

The **MyCompanyEventSource** class also includes a static member called **Log** that provides access to a shared instance of the **MyCompanyEventSource** class, and defines the constructor as private. The following code sample shows an alternative approach for defining the access to the singleton.

```
private static readonly Lazy<MyCompanyEventSource> Instance =
          new Lazy<MyCompanyEventSource>(() => new MyCompanyEventSource());

private MyCompanyEventSource() { }

public static MyCompanyEventSource Log { get { return Instance.Value; } }
```

Specifying the Event and its Payload

The event methods in your event source class enable you to provide discrete pieces of information about the event to include in the payload. For example, the **PageStart** event method enables you to include a URL in the payload. Some sinks, will store payload items individually; for example, the Windows Azure Table storage sink uses a separate column for each payload item. Payload items don't necessarily have to be strings.

Developers in the team typically create a method in the custom event source class with the payload items and the bare minimum relevant metadata if there is any, and get back to coding the business logic. At a later point, someone on the team looks at the event source class as a whole and starts assigning the appropriate metadata to make the output of the entries consistent with each other.

The **Event** attribute can contain additional metadata parameters to further refine the output, and every parameter in this attribute is optional. The only required value is the event ID.

An event's message is a human readable version of the payload information whose format you can control using the **Event** attribute's **Message** parameter. The **Startup** method always writes the string "Starting up" to the log when you invoke it. The **PageStart** method writes a message to the log, substituting the values of the **ID** and **url** parameters for the two placeholders in the string "loading page {1} activityID={0}."

> *Although the event source class does a lot of the heavy lifting for you, you still need to do your part. For example, you must also pass these parameters on to the call to the **WriteEvent** method, and ensure that you pass them in the same order as you define them in the method signature. However, this is one of the checks that the **EventSourceAnalyzer** class can perform for you. Usage of this class is described later in this chapter.*

Specifying the Log Level

You can use the **Level** parameter of the **Event** attribute to specify the severity level of the message. The **EventLevel** enumeration determines the available log levels: **Verbose (5)**, **Informational (4)**, **Warning (3)**, **Error (2)**, **Critical (1)**, and **LogAlways (0)**. **Informational** is the default logging level when not specified. When you enable an event source in your application, you can specify a log level, and the event source will log all log messages with same or lower log level. For example, if you enable an event source with the warning log level, all log methods with a level parameter value of **Warning**, **Error**, **Critical**, and **LogAlways** will be able to write log messages.

You should use log levels less than **Informational** for relatively rare warnings or errors. When in doubt stick with the default of **Informational** level, and use the **Verbose** level for events that can happen more than 1,000 times per second. Typically, users filter by severity level first and then, if necessary, refine the filter using keywords.

Using Keywords

The example includes a **Keywords** parameter for the **Event** attribute for many of the log methods. You can use the optional keywords to define different groups of events so that when you enable an event source, you can specify which groups of events to enable: only events whose **Keywords** parameter matches one of the specified groups will be able to write to the log. You can also use the keywords to filter and analyze the events in your log.

If you decide to use keywords, you must define the keywords you will use in a nested class called **Keywords** as shown in the example. Each keyword value is a 64-bit integer, which is treated as a bit array enabling you to define up to 64 different keywords. You can associate a log method with multiple keywords as shown in the following example where the **Failure** message is associated with both the **Diagnostic** and **Perf** keywords.

Although **Keywords** looks like an enumeration, it's a static class with constants of type **System.Diagnostics.Tracing.EventKeywords.** But just as with flags, you need to make sure you assign powers of two as the value for each constant.

```
[Event(1, Message = "Application Failure: {0}", Level = EventLevel.Critical,
          Keywords = Keywords.Diagnostic|Keywords.Perf)]
internal void Failure(string message)
{
  if (this.IsEnabled()) this.WriteEvent(1, message);
}
```

The following list offers some recommendations for using keywords in your organization.

- Events that you expect to fire less than 100 times per second do not need special treatment. You should use a default keyword for these events.
- Events that you expect to fire more than 1000 times per second should have a keyword. This will enable you to turn them off if you don't need them.
- It's up to you to decide whether events that typically fire at a frequency between these two values should have a keyword.
- Users will typically want to switch on a specific keyword when they enable an event source, or enable all keywords.
- Even when the frequency of events is not high, you might want to use keywords to be able to analyze the log and filter messages at a later point.

Keep it simple for users to enable just the events they need.

The logs contain just the numeric task and opcode identifiers. The developers who write the **EventSource** class and IT Pros who use the logs must agree on the definitions of the tasks and opcodes used in the application. Notice how in the sample, the task constants have meaningful names such as **Page** and **DBQuery**: these tasks appear in the logs as task ids **1** and **2** respectively.

Using Opcodes and Tasks

You can use the **Opcodes** and **Tasks** parameters of the **Event** attribute to add additional information to the message that the event source logs. The **Opcodes** and **Tasks** are defined using nested classes of the same name in a similar way to how you define **Keywords: Opcodes** and **Tasks** don't need to be assigned values that are powers of two. The example event source includes two tasks: **Page** and **DBQuery**.

*If you choose to define custom opcodes, you should assign integer values of 11 or above, otherwise they will clash with the opcodes defined in the **EventOpcode** enumeration. If you define a custom opcode with a value of 10 or below, messages that use these opcodes will not be delivered.*

Sensitive Data

You should make sure that you do not write sensitive data to the logs where it may be available to someone who shouldn't have access to that data. One possible approach is to scrub sensitive data from the event in the **EventSource** class itself. For example, if you had a requirement to write details of a connection string in a log event, you could remove the password using the following technique.

```
[Event(250, Level = EventLevel.Informational,
      Keywords = Keywords.DataAccess, Task = Tasks.Initialize)]
public void ExpenseRepositoryInitialized(string connectionString)
{
  if (this.IsEnabled(EventLevel.Informational, Keywords.DataAccess))
  {
    // Remove sensitive data
    var csb = new SqlConnectionStringBuilder(connectionString)
                { Password = string.Empty };
    this.WriteEvent(250, csb.ConnectionString);
  }
}
```

Verifying your EventSource Class

As you've seen, there are a number of conventions that you must follow when you author a custom **EventSource** class, such as matching the id in the **Event** attribute to the id passed to the **WriteEvent** method that aren't checked by the standard tools in Visual Studio or by the **EventSource** class at run time. Although the **EventSource** class does perform some basic checks, such as that the supplied event ID is valid, these checks are not exhaustive. Using the Semantic Logging Application Block, you can check that your custom event source for such common errors as part of your unit testing. The Semantic Logging Application Block includes a helper class for this purpose named **EventSourceAnalyzer**. The **Inspect** method checks your **EventSource** class as shown in the following code sample.

```
[TestMethod]
public void ShouldValidateEventSource()
{
    EventSourceAnalyzer.InspectAll(SemanticLoggingEventSource.Log);
}
```

If your Visual Studio solution does not include test projects, you can still use the **Inspect** method to check any custom **EventSource** class as shown in the sample code that accompanies this guide.

The **EventSourceAnalyzer** class checks your custom **EventSource** class using the following techniques.

- It attempts to enable a listener using the custom **EventSource** class to check for basic problems.
- It attempts to retrieve the event schemas by generating a manifest from the custom **EventSource** class.
- It attempts to invoke each event method in the custom **EventSource** class to check that it returns without error, supplies the correct event ID, and that all the payload parameters are passed in the correct order.

For more information about the **EventSourceAnalyzer** class, see the topic *Checking an EventSource Class for Errors* in the Enterprise Library Reference Documentation.

Versioning your EventSource Class

The methods in your custom **EventSource** class will be called from multiple places in your application, and possibly from multiple applications if you share the same **EventSource** class. You should take care when you modify your **EventSource** class, that any changes you make do not have unexpected consequences for your existing applications. If you do need to modify your **EventSource** class, you should restrict your changes to adding methods to support new log messages, and adding overloads of existing methods (that would have a new event ID). You should not delete or change the signature of existing methods.

This is especially important in light of the fact that if you have multiple versions of your **EventSource** class in multiple applications that are using the Semantic Logging Application Block out-of-process approach, then the version of the event schema that the host service uses will not be predictable. You should ensure that you have procedures in place to synchronize any changes to the **EventSource** class across all of the applications that share it.

Adding the Semantic Logging Application Block to Your Project

You don't need to add a reference to the Semantic Logging Application Block to your application if you use the block out-of-process.

Before you write any code that uses the Semantic Logging Application Block, you must install the required assemblies to your project. You can install the block by using the NuGet package manager in Visual Studio: in the **Manage Nuget Packages** dialog, search online for the **EnterpriseLibrary.SemanticLogging** package and install it. If you plan to use the database sinks you also need to add the **EnterpriseLibrary.SemanticLogging.Database** NuGet package. If you plan to use the Windows Azure Table storage sink, you also need to add the **EnterpriseLibrary.SemanticLogging.WindowsAzure** NuGet package.

Configuring the Semantic Logging Application Block

How you configure the Semantic Logging Application Block depends on whether you are using it in-process or out-of-process. If you are using the block in-process, then you provide the configuration information for your sinks in code; if you are using the block out-of-process, then you provide the configuration information in an XML file. The section "How do I Use the Semantic Logging Application Block to Log Trace Messages Out-of-Process?" later in this chapter describes the configuration in the out-of-process scenario. This section describes the in-process scenario.

Typically, you create an **ObservableEventListener** instance to receive the events from your application, then you create any sinks you need to save the log messages. The built-in sinks have constructors that enable you to configure the sink as you create it. However, the block includes convenience methods, such as **LogToConsole** and **LogToSqlDatabase**, to set up the sinks and attach them to a listener. Once the listener is created, you enable one or more event sources to listen to, as well as the highest level of event to capture, and any keywords to filter on for each source. The following code sample shows an example that creates, configures, and enables two sinks, one to write log messages to the console and one to write log messages to a SQL Database.

```
// Initialize the listeners and sinks during application start-up
var listener1 = new ObservableEventListener();

listener1.EnableEvents(
  MyCompanyEventSource.Log, EventLevel.LogAlways,
  MyCompanyEventSource.Keywords.Perf |
  MyCompanyEventSource.Keywords.Diagnostic);

listener1.LogToConsole();

var listener2 = new ObservableEventListener();
```

```
listener2.EnableEvents(
  MyCompanyEventSource.Log, EventLevel.LogAlways, Keywords.All);

// The SinkSubscription is used later to flush the buffer,
// although this is typically not needed for basic scenarios
SinkSubscription<SqlDatabaseSink> subscription =
  listener2.LogToSqlDatabase("Demo Semantic Logging Instance",
  connectionString);
```

This example uses two listeners because it is using different keyword filters for each one.

An alternative approach is to use the static entry points provided by the log classes as shown in the following code sample. The advantage of this approach is you only need to invoke a single method rather than creating the listener and attaching the sink. However, this method does not give you access to the sink and therefore you cannot explicitly flush any buffers associated with the sink when you are shutting down the application.

You should realize that creating an instance of an observable event listener results in shared state. You should set up your listeners at bootstrap time and dispose of them when you shut down your application.

You can use the **onCompletedTimeout** parameter to the **CreateListener** method controls how long a listener will wait for the sink to flush itself before disposing the sinks. For more information, see the topic *Event Sink Properties* in the Enterprise Library Reference Documentation.

```
// Create the event listener using the static method.
// Typically done when the application starts.
var listener = ConsoleLog.CreateListener();
listener.EnableEvents(MyCompanyEventSource.Log, EventLevel.LogAlways,
                      Keywords.All);
...
// Disable and dispose the event listener.
// Typically done when the application terminates.
listener.DisableEvents(MyCompanyEventSource.Log);
listener.Dispose();
```

The console and file based sinks can also use a formatter to control the format of the output. These formatters may also have configuration options. The following code sample shows an example that configures a JSON formatter for the console sink, and uses a custom console color mapper.

```
JsonEventTextFormatter formatter =
      new JsonEventTextFormatter(EventTextFormatting.Indented);

var colorMapper = new MyCustomColorMapper();
listener1.LogToConsole(formatter, colorMapper);
listener1.EnableEvents(
      MyCompanyEventSource.Log, EventLevel.LogAlways, Keywords.All);
```

A color mapper is a simple class that specifies the color to use for different event levels as shown in the following code sample.

```
public class MyCustomColorMapper : IconsoleColorMapper
{
  public ConsoleColor? Map(
    System.Diagnostics.Tracing.EventLevel eventLevel)
  {
    switch (eventLevel)
    {
      case EventLevel.Critical:
        return ConsoleColor.White;
      case EventLevel.Error:
        return ConsoleColor.DarkMagenta;
      case EventLevel.Warning:
        return ConsoleColor.DarkYellow;
      case EventLevel.Verbose:
        return ConsoleColor.Blue;
      default:
        return null;
    }
  }
```

Writing to the Log

Before you can write a log message, you must have an event source class in your application that defines what log messages you can write. The section "Creating an Event Source" earlier in this chapter describes how you can define such an event source. You must also add the Semantic Logging Application Block to your application: the easiest way to do this is using NuGet.

The following code sample shows a simple example of how you can use the event source shown previously in this chapter to write messages to a console window.

```
class Program
{
  static void Main(string[] args)
  {
    // Create the event listener
    var listener = new ObservableEventListener();
    listener.EnableEvents(MyCompanyEventSource.Log, EventLevel.LogAlways,
                          Keywords.All);
    listener.LogToConsole();

    MyCompanyEventSource.Log.StartUp();
    // Application code goes here
    ...

    listener.DisableEvents(MyCompanyEventSource.Log);
    listener.Dispose();
  }
}
```

This sample shows how to create an instance of the **ObservableEventListener** class from the Semantic Logging Application Block, enable the listener to process log messages from the **MyCompanyEventSource** class, write a log message, disable the listener, and then dispose of the listener. Typically, you will create and enable an event listener when your application starts up and initializes, and disable and dispose of your event listener as the application shuts down.

When you enable an event listener for an event source, you can specify which level of events from that event source should be logged and which groups of events (identified using the keywords) should be active. The example shows how to activate log methods at all levels and with all keywords.

> *If you want to activate events in all groups, you must use the **Keywords.All** parameter value. If you do not supply a value for this optional parameter, only events with no keywords are active.*

The following code sample shows how to activate log messages with a level of **Warning** or lower with keywords of either **Perf** or **Diagnostic**.

```
listener.EnableEvents(MyCompanyEventSource.Log, EventLevel.Warning,
  MyCompanyEventSource.Keywords.Perf |
  MyCompanyEventSource.Keywords.Diagnostic);
```

How Do I Use the Semantic Logging Application Block to Log Events Out-of-Process?

To use the Semantic Logging Application Block to process events out-of-process you must add the necessary code to your LOB application to create trace messages when interesting events occur in the application in the same way that you do in-process, by creating a class derived from **EventSource**. You must also run and configure a separate process that is responsible for collecting and processing the events.

The LOB application itself will use the built-in **EventSource** infrastructure to write events through ETW, and will not require a reference to the Semantic Logging Application Block.

Running the Out-of-Process Host Application

The Semantic Logging Application Block includes the SemanticLogging-svc host application for you to use; you can run this application as a Windows Service or as a console application.

Typically, you should run the Out-of-Process Host as a console application when you are developing and testing your logging behavior. It's convenient to be able to stop and start the event listener host application and view any log messages in a console window.

In a production environment, you should run the Out-of-Process Host as a Windows Service. You can easily configure a Windows Service to start when the operating system starts. In a production environment, you are unlikely to want to see log messages as they are processed in a console window: more likely, you will want to save the log messages to a file, database, or some other persistent storage.

For those sinks that buffer log messages, such as the Windows Azure Table storage sink, you can flush the sink by calling the **FlushAsync** method to ensure that all events are written. Nevertheless, the sink will also flush automatically when disposing the listener. Keep in mind that the **Dispose** call will block the thread until the events are persisted, but you can also specify a timeout when initializing the sink.

You should consider collecting and processing trace messages in a separate process to improve the robustness of your logging solution. If your LOB crashes, a separate process will be responsible for any logging.

You should also consider collecting and processing events in a separate process if you are using a sink with a high latency such as the Windows Azure Table storage sink and you want the buffering to not prevent the application from shutting down very quickly.

As an alternative to the Out-of-Process Host application included with the block, you could configure ETW to save the trace messages to a .etl file and use another tool to read this log file. However, the Out-of-Process Host application offers more choices of where to persist the log messages.

There may be a limit on the number of ETW sessions that you can create on a machine dependent on the resources available on that machine. The block will log an error if it can't create sessions for all of the sinks defined in your SemanticLogging-svc.xml file.

Each **EventSource** implementation should have a unique name, although you might choose to reuse an **EventSource** in multiple applications. If you use the same name for the event source in multiple applications, all of the events from those applications will be collected and processed by a single logging application. If you omit the **EventSource** attribute, ETW uses the class name as the name of the event source.

If you plan to use an out-of-process host in Windows Azure, you should run it as a Windows Service. You can install and start the service in a Windows Azure start up task.

By default, the Out-of-Process Host application reads the configuration for the block from a file named SemanticLogging-svc.xml. It reads this file at start up, and then continues to monitor this file for changes. If it detects any changes, it dynamically reconfigures the block based in the changes it discovers. If you make any changes to the **traceEventService** element, this will cause the service to recycle; for any changes related to sink definitions, the block will reconfigure itself without recycling the service. If the block detects any errors when it loads the new configuration, it will log an error in the Windows Event Log and continue to process the remainder of the configuration file. You should check the messages in the Windows Event Log to verify that your changes were successful.

Updating an Event Source Class in the Out-of-Process Scenario

You should be careful if you use the same custom **EventSource** class in multiple applications. It's possible in the out-of-process scenario that you might modify this **EventSource** class in one of the applications that's creating event messages, and if you don't change the event source name or id using the attributes in the event source class, the changes will be picked up by the out-of-process host without it restarting. The most recent changes to the custom **EventSource** class will take precedence and be used by the out-of-process host, which may have an impact on the logging behavior of the other applications. Ideally, you should keep your event source class in sync across any applications that share it.

Creating Trace Messages

You create trace messages in the LOB application in exactly the same way as described previously in this chapter. First, create a custom **EventSource** class that defines all of the events your application uses as shown in the following code sample.

```
[EventSource(Name = "MyCompany")]
public class MyCompanyEventSource : EventSource
{
    ...

    [Event(1, Message = "Application Failure: {0}",
           Level = EventLevel.Critical, Keywords = Keywords.Diagnostic)]
    internal void Failure(string message)
    {
      if (this.IsEnabled()) this.WriteEvent(1, message);
    }

    ...

    public static readonly MyCompanyEventSource Log =
                                               new MyCompanyEventSource();
}
```

Second, create the trace messages in your application code as shown in the following code sample.

```
MyCompanyEventSource.Log.CouldNotConnectToServer();
```

You don't need to create any event listeners or sinks in the LOB application if you are processing the log messages in a separate application.

Choosing Sinks

The Semantic Logging Application Block includes sinks that enable you to save log messages to the following locations: a database, a Windows Azure table, and to flat files. The choice of which sinks to use depends on how you plan to analyze and monitor the information you are collecting. For example, if you save your log messages to a database you can use SQL to query and analyze the data. If you are collecting log messages in a Windows Azure application you should consider writing the messages to Windows Azure table storage. You can export data from Windows Azure Table storage to a local file for further analysis on-premises.

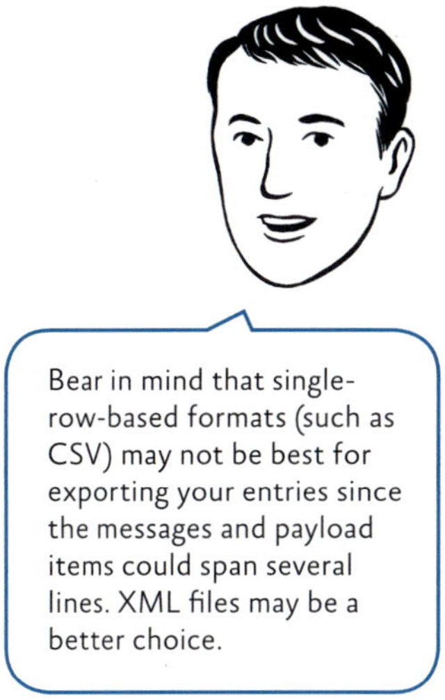

The console sink is useful during development and testing because you can easily see the trace messages as they appear, however the console sink is not appropriate for use in a production environment.

If you use the ***XmlEventTextFormatter*** *class to save the log messages formatted as XML in a flat file, you will need to perform some post-processing on the log files before you can use a tool such as Log Parser to read them. This is because the flat file does not have an XML root element, therefore the file does not contain well-formed XML. It's easy to add opening and closing tags for a root element to the start and end of the file.*

Although the block does not include a CSV formatter, you can easily export log messages in CSV format from Windows Azure Table storage or a SQL Server database.

The Log Parser tool can add a header to some file formats such as CSV, but it cannot add a header or a footer to an XML file.

Collecting and Processing the Log Messages

To collect and process the log messages from the LOB application you run the Enterprise Library Semantic Logging Out-of-Process Windows Service/Console Host (SemanticLogging-svc.exe) included with the Semantic Logging Application Block. You can run this application as a console application or install it as a Windows service. This application uses configuration information from an XML file to determine which event sources to collect trace messages from and which sinks to use to process those messages.

The default name for the configuration file is SemanticLogging-svc.xml. When you edit this XML file in Visual Studio, you can use the supplied schema file (SemanticLogging-svc.xsd) to provide IntelliSense support in the editor.

The following snippet shows an example configuration file that defines how to collect trace messages written by another application using the Adatum event source. This example uses three sinks to write log messages to three different destinations. Each sink is configured to log messages with different severity levels. Additionally, the console sink filters for messages that have the **Diagnostics** or **Perf** keywords (The **Diagnostics** constant has a value of four and the Perf constant has a value of eight in the nested **Keywords** class in the **AdatumEvent-Source** class).

```
<?xml version="1.0"?>
<configuration
  xmlns="http://schemas.microsoft.com/practices/2013/entlib/semanticlogging/etw"
  xmlns:xsi="http://www.w3.org/2001/XMLSchema-instance"
  xsi:schemaLocation="http://schemas.microsoft.com/practices/2013/entlib/
  semanticlogging/etw SemanticLogging-svc.xsd">
  <!-- Optional settings for this host -->
  <traceEventService/>

  <!-- Event sink definitions used by this host to
  listen ETW events emitted by these EventSource instances -->
  <sinks>
    <consoleSink name="ConsoleEventSink">
      <sources>
        <eventSource name="MyCompany" level="LogAlways" matchAnyKeyword="12" />
      </sources>
      <eventTextFormatter header="+=========================================+" />
    </consoleSink>

    <rollingFlatFileSink name="RollingFlatFileSink"
                         fileName="RollingFlatFile.log"
                         timeStampPattern="yyyy"
                         rollFileExistsBehavior="Overwrite"
                         rollInterval="Day">
      <sources>
        <eventSource name="MyCompany"
                     level="Error" />
      </sources>
    </rollingFlatFileSink>

    <sqlDatabaseSink name="SQL Database Sink"
                     instanceName="Demo"
                     connectionString="Data Source=(localDB)\v11.0;
                                       Initial Catalog=Logging;
                                       Integrated Security=True">
      <sources>
        <eventSource name="MyCompany"
                     level="Warning" />
      </sources>
    </sqlDatabaseSink>
  </sinks>

</configuration>
```

If you are collecting high volumes of trace messages from a production application, you may need to tweak the settings for the trace event service in the XML configuration file. For information about these settings, see the topic *Performance Considerations* in the Enterprise Library Reference Documentation.

A single instance of the console application or Windows service can collect messages from multiple event sources in multiple LOB applications. You can also run multiple instances, but each instance must use a unique session name.

By default, each instance of the console application has a unique session name. However, if you decide to provide a session name in the XML configuration file as part of the trace event service settings, you must ensure that the name you choose is unique.

Customizing the Semantic Logging Application Block

The Semantic Logging Application Block provides a number of extension points that enable you to further customize its behavior. These extension points include:

- Creating custom filters and manipulating events using Reactive Extensions.
- Creating custom formatters for use both in-process and out-of-process.
- Creating custom sinks for use both in-process and out-of-process.
- Creating a custom application to collect trace messages from your LOB applications.

The example code in this section comes from the sample applications that are included with this guidance. In some cases, the code has been changed slightly to make it easier to read.

Creating Custom Filters Using Reactive Extensions (Rx)

The sink classes included with the Semantic Logging Application Block implement the **IObserver** interface, which enables them to subscribe to event listeners that implement the **IObservable** interface. By creating custom code that subscribes to an **IObservable** instance and that generates a new stream that an **IObserver** instance can subscribe to, you can manipulate the events before they are seen by the sink. This example illustrates how you can use Reactive Extensions to build a custom filter that controls which events are sent to the sinks in the in-process scenarios.

The Semantic Logging Application Block does not have a dependency on Reactive Extensions (Rx), but this specific example does require you to add the Rx NuGet package to your project because it uses the **Observable** class and the utilities that the Rx library brings.

This example shows how you can buffer some log messages until a specific event takes place. In this example, no informational messages are sent to the sink, until the listener receives an error message. When this happens, the listener the sends the last ten buffered informational messages and the error message to the sink. This is useful if, for the majority of the time you are not interested in the informational messages, but when an error occurs, you want to see the most recent informational messages that might have led to that error.

The following code sample shows the **FlushOnTrigger** extension method that implements this behavior. It uses a class called **CircularBuffer** to hold the most recent messages, and the sample application includes a simple implementation of this **CircularBuffer** class.

```
public static IObservable<T> FlushOnTrigger<T>(
                                this IObservable<T> stream, Func<T, bool> shouldFlush, int bufferSize)
{
  return Observable.Create<T>(observer =>
  {
    var buffer = new CircularBuffer<T>(bufferSize);
    var subscription = stream.Subscribe(newItem =>
      {
        if (shouldFlush(newItem))
        {
          foreach (var buffered in buffer.TakeAll())
          {
            observer.OnNext(buffered);
          }

          observer.OnNext(newItem);
        }
        else
        {
          buffer.Add(newItem);
        }
      },
      observer.OnError,
      observer.OnCompleted);

    return subscription;
  });
}
```

The following code sample shows how to use this method when you are initializing a sink. In this scenario, you must invoke the **FlushOnTrigger** extension method that creates the new stream of events for the sink to subscribe to.

```
var listener = new ObservableEventListener();
listener.EnableEvents(MyCompanyEventSource.Log, EventLevel.Informational,
                      Keywords.All);

listener
  .FlushOnTrigger(entry => entry.Schema.Level <= EventLevel.Error, bufferSize: 10)
  .LogToConsole();
```

Creating Custom Formatters

The Semantic Logging Application Block includes three text formatters to format the log messages written to the console or to text files: **EventTextFormatter**, **JsonTextFormatter**, and **XmlTextFormatter**.

There are three different scenarios to consider if you are planning to create a new custom event text formatter for the Semantic Logging Application Block. They are listed in order of increasing level of complexity to implement:

- Creating a custom formatter for use in-process.
- Creating a custom formatter for use out-of-process without IntelliSense support when you edit the configuration file.
- Creating a custom formatter for use out-of-process with IntelliSense support when you edit the configuration file.

Creating a Custom In-Process Event Text Formatter

To create a custom text formatter, you create a class that implements the **IEventTextFormatter** interface and then use that class as your custom formatter. This interface contains a single method called **WriteEvent**.

The following code sample shows how to create a custom text formatter similar to the built-in **EventText-Formatter**, but prefixes the properties of an event entry with a user supplied value.

```
public class PrefixEventTextFormatter : IeventTextFormatter
{
  public PrefixEventTextFormatter(string header, string footer,
                                string prefix, string dateTimeFormat)
  {
    this.Header = header;
    this.Footer = footer;
    this.Prefix = prefix;
    this.DateTimeFormat = dateTimeFormat;
  }

  public string Header { get; set; }
  public string Footer { get; set; }
  public string Prefix  { get; set; }
  public string DateTimeFormat { get; set; }

  public void WriteEvent(EventEntry eventEntry, TextWriter writer)
  {
    // Write header
    if (!string.IsNullOrWhiteSpace(this.Header))
        writer.WriteLine(this.Header);

    // Write properties
    writer.WriteLine("{0}SourceId : {1}",
                     this.Prefix, eventEntry.ProviderId);
    writer.WriteLine("{0}EventId : {1}",
                     this.Prefix, eventEntry.EventId);
    writer.WriteLine("{0}Keywords : {1}",
                     this.Prefix, eventEntry.Schema.Keywords);
    writer.WriteLine("{0}Level : {1}",
                     this.Prefix, eventEntry.Schema.Level);
    writer.WriteLine("{0}Message : {1}",
                     this.Prefix, eventEntry.FormattedMessage);
    writer.WriteLine("{0}Opcode : {1}",
                     this.Prefix, eventEntry.Schema.Opcode);
    writer.WriteLine("{0}Task : {1} {2}",
                     this.Prefix, eventEntry.Schema.Task, eventEntry.Schema.TaskName);
    writer.WriteLine("{0}Version : {1}",
                     this.Prefix, eventEntry.Schema.Version);
    writer.WriteLine("{0}Payload :{1}",
                     this.Prefix, FormatPayload(eventEntry));
    writer.WriteLine("{0}Timestamp : {1}",
                     this.Prefix, eventEntry.GetFormattedTimestamp(this.DateTimeFormat));
```

```
        // Write footer
        if (!string.IsNullOrWhiteSpace(this.Footer))
            writer.WriteLine(this.Footer);

        writer.WriteLine();
    }

    private static string FormatPayload(EventEntry entry)
    {
        var eventSchema = entry.Schema;
        var sb = new StringBuilder();
        for (int i = 0; i < entry.Payload.Count; i++)
        {
            // Any errors will be handled in the sink
            sb.AppendFormat(" [{0} : {1}]",
                            eventSchema.Payload[i], entry.Payload[i]);
        }
        return sb.ToString();
    }
}
```

You can use the **CustomFormatterUnhandledFault** method in the **SemanticLoggingEventSource** class to log any unhandled exceptions in your custom formatter class, but you should always avoid throwing an exception in the **WriteEvent** method body.

Using text formatters when you are using the Semantic Logging Application Block in-process is a simple case of instantiating and configuring in code the formatter you want to use, and then passing it as a parameter to the method that initializes the sink that you're using. The following code sample shows how to create and configure an instance of the **PrefixEventTextFormatter** formatter class and pass it to the **LogToConsole** extension method that subscribes **ConsoleSink** instance to the event listener:

```
var formatter = new PrefixEventTextFormatter("---", null, ">> ", "d");

listener.LogToConsole(formatter);
listener.EnableEvents(MyCompanyEventSource.Log, EventLevel.LogAlways, Keywords.All);
```

Creating a Custom Out-of-Process Event Text Formatter without IntelliSense Support

Built-in formatters have IntelliSense support in the XML for configuring the Out-of-Process host, but custom formatters do not by default. You can use the custom event text formatter shown in the previous section in an out-of-process scenario by using the **customEventTextFormatter** element in the configuration file. The following XML sample shows how to use this built-in element.

```
<sinks>
  <consoleSink name="Consolesink">
    <sources>
      <eventSource name="MyCompany" level="LogAlways" />
    </sources>

    <customEventTextFormatter
      type="CustomTextFormatter.PrefixEventTextFormatter, CustomTextFormatter">
      <parameters>
        <parameter name="header" type="System.String"
          value="==============================================" />
        <parameter name="footer" type="System.String"
          value="==============================================" />
        <parameter name="prefix" type="System.String" value="> " />
        <parameter name="dateTimeFormat" type="System.String" value="O" />
      </parameters>
    </customEventTextFormatter>
  </consoleSink>
  ...
```

This example illustrates how you must specify the type of the custom formatter, and use **parameter** elements to define the configuration settings. The out-of-process host application looks scans the folder it runs from for DLLs that contain types that implement the **IEventTextFormatter** interface.

You can use the same class that implements the IEventTextFormatter interface in both in-process and out-of-process scenarios.

You must ensure that the order and type of the **parameter** elements in the configuration file match the order and type of the constructor parameters in the custom formatter class.

Creating a Custom Out-of-Process Event Text Formatter with IntelliSense Support

Instead of using the **customEventTextFormatter** and **parameter** elements in the XML configuration file, you can define your own custom element and attributes and enable IntelliSense support in the Visual Studio XML editor. In this scenario, the XML configuration file looks like the following sample:

```
<?xml version="1.0"?>
<configuration
   xmlns=http://schemas.microsoft.com/practices/2013/entlib/semanticlogging/etw
   xmlns:xsi=http://www.w3.org/2001/XMLSchema-instance
   xsi:schemaLocation="urn:sample.etw.customformatter
   PrefixEventTextFormatterElement.xsd
   http://schemas.microsoft.com/practices/2013/entlib/semanticlogging/etw
   SemanticLogging-svc.xsd">
<traceEventService />

<sinks>
  <consoleSink name="Consolesink">
    <sources>
      <eventSource name="MyCompany" level="LogAlways" />
    </sources>
```

Notice how the **prefixEventTextFormatter** element has a custom XML namespace and uses a **schemaLocation** to specify the location of the schema file.

```
    <prefixEventTextFormatter xmlns="urn:sample.etw.customformatter"
     header="================================================================="
     footer="================================================================="
     prefix="# "
     dateTimeFormat="O" />
  </consoleSink>
  ...
```

The custom XML namespace that enables IntelliSense behavior for the contents of the **prefixEventTextFormatter** element is defined in the XML schema file shown in the following sample.

```
<?xml version="1.0" encoding="utf-8"?>
<xs:schema id="PrefixEventTextFormatterElement"
    targetNamespace="urn:sample.etw.customformatter"
    xmlns="urn:sample.etw.customformatter"
    xmlns:xs="http://www.w3.org/2001/XMLSchema"
    elementFormDefault="qualified"
    attributeFormDefault="unqualified">

  <xs:element name="prefixEventTextFormatter">
    <xs:complexType>
      <xs:attribute name="header" type="xs:string" use="required" />
      <xs:attribute name="footer" type="xs:string" use="required" />
      <xs:attribute name="prefix" type="xs:string" use="required" />
      <xs:attribute name="dateTimeFormat" type="xs:string" use="required" />
    </xs:complexType>
  </xs:element>
</xs:schema>
```

You should place this schema file, along with the XML configuration file, in the folder where you are running the listener host application.

In addition to creating the XML schema, you must also create an element definition that enables the block to read the configuration from the XML file and create an instance of the formatter. The following code sample shows the element definition class for this example.

```
public class PrefixEventTextFormatterElement : IformatterElement
{
  private readonly XName formatterName =
    XName.Get("prefixEventTextFormatter", "urn:sample.etw.customformatter");

  public bool CanCreateFormatter(System.Xml.Linq.XElement element)
  {
    return this.GetFormatterElement(element) != null;
  }
```

```
  public IEventTextFormatter CreateFormatter(System.Xml.Linq.XElement element)
  {
    var formatter = this.GetFormatterElement(element);

    var header = (string)formatter.Attribute("header");
    var footer = (string)formatter.Attribute("footer");
    var prefix = (string)formatter.Attribute("prefix");
    var datetimeFormat = (string)formatter.Attribute("dateTimeFormat");

    return new PrefixEventTextFormatter(header,footer, prefix, datetimeFormat);
  }

  private XElement GetFormatterElement(XElement element)
  {
    return element.Element(this.formatterName);
  }
}
```

This class implements the **IFormatterElement** interface that includes the methods **CanCreateFormatter** and **CreateFormatter**.

Creating Custom Sinks

The Semantic Logging Application Block includes a number of sinks for receiving and processing log messages such as the **ConsoleSink**, the **EventLogSink**, and the **RollingFlatFileSink**.

There are three different scenarios to consider if you are planning to create a new custom sink for the Semantic Logging Application Block. They are listed in order of increasing level of complexity to implement:

- Creating a custom sink for use in-process.
- Creating a custom sink for use out-of-process without IntelliSense support when you edit the configuration file.
- Creating a custom sink for use out-of-process with IntelliSense support when you edit the configuration file.

Creating a Custom In-Process Sink

To create a custom sink to use in-process, you must create a new sink class that implements the **IObserver<EventEntry>** interface. Typically, you implement the **OnNext** method from this interface as shown in the following example from the **EmailSink** class.

```
public sealed class EmailSink : IObserver<EventEntry>
{
  private const string DefaultSubject = "Email Sink Extension";
  private IEventTextFormatter formatter;
  private MailAddress sender;
  private MailAddressCollection recipients = new MailAddressCollection();
  private string subject;
  private string host;
  private int port;
  private NetworkCredential credentials;
```

```
public EmailSink(string host, int port,
                 string recipients, string subject, string credentials,
                 IEventTextFormatter formatter)
{
  this.formatter = formatter ?? new EventTextFormatter();
  this.host = host;
  this.port = GuardPort(port);
  this.credentials = CredentialManager.GetCredentials(credentials);
  this.sender = new MailAddress(this.credentials.UserName);
  this.recipients.Add(GuardRecipients(recipients));
  this.subject = subject ?? DefaultSubject;
}

public void OnNext(EventEntry entry)
{
  if (entry != null)
  {
    using (var writer = new StringWriter())
    {
      this.formatter.WriteEvent(entry, writer);
      Post(writer.ToString());
    }
  }
}

private async void Post(string body)
{
  using (var client = new SmtpClient(this.host, this.port)
       { Credentials = this.credentials, EnableSsl = true })
  using (var message = new MailMessage(this.sender, this.recipients[0])
       { Body = body, Subject = this.subject })
  {
    for (int i = 1; i < this.recipients.Count; i++)
       message.CC.Add(this.recipients[i]);

    try
    {
      await client.SendMailAsync(message).ConfigureAwait(false);
    }
    catch (SmtpException e)
    {
      SemanticLoggingEventSource.Log.CustomSinkUnhandledFault(
                                          "SMTP error sending email: " + e.Message);
    }
    catch (InvalidOperationException e)
    {
      SemanticLoggingEventSource.Log.CustomSinkUnhandledFault(
                                         "Configuration error sending email: " + e.Message);
    }
  }
}
```

```
    public void OnCompleted()  {
    }

    public void OnError(Exception error)  {
    }

    private static int GuardPort(int port)
    {
      if (port < 0)
        throw new ArgumentOutOfRangeException(“port”);

      return port;
    }

    private static string GuardRecipients(string recipients)
    {
      if (recipients == null)
        throw new ArgumentNullException("recipients");
      if (string.IsNullOrWhiteSpace(recipients))
        throw new ArgumentException(
                 "The recipients cannot be empty", "recipients");

      return recipients;
    }
  }
```

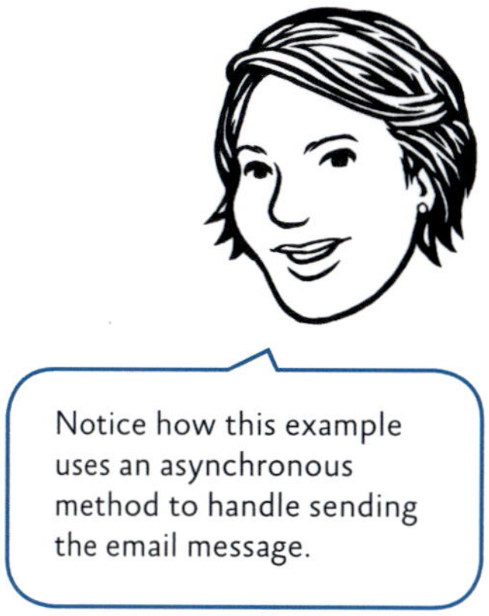

You can use the **CustomSinkUnhandledFault** method in the **SemanticLogging-EventSource** class to log exceptions in your custom sink.

To learn how to create a custom in-process sink you should also examine the source code of some of the built-in event sinks such as the **ConsoleSink** or **SQLDatabaseSink**. The **SQLDatabaseSink** illustrates an approach to buffering events in the sink.

Before you use the custom sink in-process, you should create an extension method to enable you to subscribe the sink to the observable event listener. The following code sample shows an extension method for the custom email sink.

```
public static class EmailSinkExtensions
{
  public static SinkSubscription<EmailSink> LogToEmail(
    this IObservable<EventEntry> eventStream, string host, int port,
    string recipients, string subject, string credentials,
    IEventTextFormatter formatter = null)
  {
    var sink = new EmailSink(host, port, recipients, subject,
                             credentials, formatter);

    var subscription = eventStream.Subscribe(sink);
    return new SinkSubscription<EmailSink>(subscription, sink);
  }
}
```

Finally, to use a sink in-process, you can create and configure the event listener in code as shown in the following code sample.

```
ObservableEventListener listener = new ObservableEventListener();
listener.EnableEvents(MyCompanyEventSource.Log,
                      EventLevel.LogAlways, Keywords.All);

listener.LogToConsole();
listener.LogToEmail("smtp.live.com", 587, "bill@adatum.com",
                    " In Proc Sample", "etw");
```

You can also pass a formatter to the **LogToEmail** method if you want to customize the format of the email message.

Creating a Custom Out-of-Process Sink without IntelliSense Support

When you use the Semantic Logging Application Block out-of-process, the block creates and configures sink instances based on configuration information in an XML file. You can configure the block to use a custom sink by using the **customSink** element as shown in the following XML sample.

```
<customSink name="SimpleCustomEmailSink"
            type ="CustomSinkExtension.EmailSink, CustomSinkExtension">
  <sources>
    <eventSource name="MyCompany" level="Critical" />
  </sources>
  <parameters>
    <parameter name="host" type="System.String" value="smtp.live.com" />
    <parameter name="port" type="System.Int32" value="587" />
    <parameter name="recipients" type="System.String"
               value="bill@adatum.com" />
    <parameter name="subject" type="System.String"
               value="Simple Custom Email Sink" />
    <parameter name="credentials" type="System.String" value="etw" />
  </parameters>
</customSink>
```

In this example, the class named **EmailSink** defines the custom sink. The **parameter** elements specify the constructor arguments in the same order that they appear in the constructor.

You can use the same custom **EmailSink** class (shown in the previous section) that implements the **IObserver<EventEntry>** interface in both in-process and out-of-process scenarios.

If you are using the **customSink** element in the XML configuration file, the order of the **parameter** elements must match the order of the constructor arguments. Your custom sink class must also implement the **IObserver<EventEntry>** interface.

Creating a Custom Out-of-Process Sink with IntelliSense Support

To make it easier to configure the custom sink, you can add support for IntelliSense when you edit the XML configuration file. You can use the same **SimpleCustomEmailSink** class shown in the previous section.

The following XML sample shows the XML that configures the custom sink and that supports IntelliSense in Visual Studio.

```
<emailSink xmlns="urn:sample.etw.emailsink"
           credentials="etw" host="smtp.live.com"
           name="ExtensionEmailSink" port="587"
           recipients="bill@adatum.com" subject="Extension Email Sink">
  <sources>
    <eventSource name="MyCompany" level="Critical" />
  </sources>
</emailSink>
```

If your custom sink does not implement **IObserver<T>** where **T** is **EventEntry**, you must provide the additional configuration support shown in this section. You must also, provide a transformation from **EventEntry** to **T**: see the built-in sinks in the block for examples of how to do this.

This example uses the custom **emailSink** element in a custom XML namespace. An XML schema defines this namespace, and enables the IntelliSense support in Visual Studio. The following XML sample shows the schema.

```
<?xml version="1.0" encoding="utf-8"?>
<xs:schema id="EmailEventSinkElement"
    targetNamespace="urn:sample.etw.emailsink"
    xmlns="urn:sample.etw.emailsink"
    xmlns:etw=
    http://schemas.microsoft.com/practices/2013/entlib/semanticlogging/etw
    xmlns:xs="http://www.w3.org/2001/XMLSchema"
    elementFormDefault="qualified"
    attributeFormDefault="unqualified">

  <xs:element name="emailSink">
    <xs:complexType>
      <xs:sequence>
        <xs:any minOccurs="0" maxOccurs="unbounded" processContents="skip"/>
      </xs:sequence>
      <xs:attribute name="name" type="xs:string" use="required" />
      <xs:attribute name="host" type="xs:string" use="required" />
      <xs:attribute name="port" type="xs:int" use="required" />
      <xs:attribute name="credentials" type="xs:string" use="required" />
      <xs:attribute name="recipients" type="xs:string" use="required" />
      <xs:attribute name="subject" type="xs:string" use=”optional" />
    </xs:complexType>
  </xs:element>
</xs:schema>
```

You should place this schema file in the same folder as your XML configuration file (or in a subfolder beneath the folder that holds your XML configuration file).

A schema file for your custom sink is optional. However, your custom sink element should be in a custom namespace to differentiate it from the built-in namespace. For example:

```
<emailSink xmlns:noxsd="urn:no_schema"
  ... >
  <sources>
    ...
  </sources>
</emailSink>
```

In addition to the schema file, you also need a class to enable the Semantic Logging Application Block to read the custom configuration data from the XML configuration file. This **EmailSinkElement** class must implement the **ISinkElement** class and implement the **CanCreateSink** and **CreateSink** methods as shown in the following example.

```
public class EmailSinkElement : IsinkElement
{
  private readonly XName sinkName = XName.Get("emailSink", "urn:sample.etw.emailsink");

  public bool CanCreateSink(XElement element)
  {
    return element.Name == this.sinkName;
  }

  public IObserver<EventEntry> CreateSink(XElement element)
  {
    var host = (string)element.Attribute("host");
    var port = (int)element.Attribute("port");
    var recipients = (string)element.Attribute("recipients");
    var subject = (string)element.Attribute("subject");
    var credentials = (string)element.Attribute("credentials");

    var formatter = FormatterElementFactory.Get(element);
    return new EmailSink(host, port, recipients, subject, credentials, formatter);
  }
}
```

The **CreateSink** method creates an instance of the custom **EmailSink** type.

Place the assembly that contains your implementations of the **ISinkElement** and **IObserver<EventEntry>** interfaces is the same folder as the XML configuration file.

Creating Custom Event Listener Host Applications

The Semantic Logging Application Block includes an event listener host application to use when you want to use the block out-of-process: this application can run as a console application or as a Windows Service. Internally, both of these applications use the **TraceEventService** and **TraceEventServiceConfiguration** classes.

You can create your own custom out-of-process event listener using these same two classes. The following code sample from the out-of-process console host illustrates their use.

```
internal class TraceEventServiceHost
{
  private const string LoggingEventSourceName = "Logging";
  private const string EtwConfigurationFileName = "EtwConfigurationFileName";

  private static readonly TraceSource logSource = new TraceSource(LoggingEventSourceName);

  internal static void Run()
  {
    try
    {
      ...
      var configuration = TraceEventServiceConfiguration.Load(
                                            ConfigurationManager.AppSettings[EtwConfigurationFileName]);
      using (var service = new TraceEventService(configuration))
      {
        service.Start();
        ...
      }
    }
    catch (Exception e)
    {   ...   }
    finally
    {
      logSource.Close();
    }
  }
  ...
}
```

You should consider adding a mechanism that monitors the configuration file for changes. The Out-of-Process Host application included with the block uses the **FileSystemWatcher** class from the .NET Framework to monitor the configuration file for changes. When it detects a change it automatically restarts the internal **TraceEventService** instance with the new configuration settings.

Summary

The Semantic Logging Application Block, while superficially similar to the Logging Application Block, offers two significant additional features.

Semantic logging and strongly typed events ensure that the structure of your log messages is well defined and consistent. This makes it much easier to process your log messages automatically whether you want to analyze them, gain operational intelligence insights, drive some other automated process from them, or correlate them with another set of events. Creating a log message in your code now becomes a simple case of calling a method to report that an event of some significance has just happened in your application. However, you do have to write the event method, but this means you make decisions about the characteristics of the message in the correct place in your application: in your event source class.

You can use the Semantic Logging Application Block out-of-process to capture and process log messages from another application. This makes it possible to collect diagnostic trace information from a live application while making logging more robust by handling log messages in a separate process. The block makes use of Event Tracing for Windows to achieve this: the production application can write high volumes of trace messages that ETW intercepts and delivers to your separate logging application. Your logging application can use all of the block's sinks and formatters to process these log messages and send them to the appropriate destination.

More Information

All links in this book are accessible from the book's online bibliography on MSDN at *http://aka.ms/el6biblio*.

For more information about ETW, see *Event Tracing* on MSDN.

For more information about the **EventSource** class and the **WriteEvent** and **IsEnabled** methods, see *EventSource Class* on MSDN.

For more information about the **EventSourceAnalyzer** class, see the topic *Checking an EventSource Class for Errors* in the Enterprise Library Reference Documentation.

You can use the **onCompletedTimeout** parameter to the **CreateListener** method controls how long a listener will wait for the sink to flush itself before disposing the sinks. For more information, see the topic *Event Sink Properties* in the Enterprise Library Reference Documentation.

If you are collecting high volumes of trace messages from a production application, you may need to tweak the settings for the trace event service in the XML configuration file. For information about these settings, see the topic *Performance Considerations* in the Enterprise Library Reference Documentation.

General Links:

- There are resources to help if you're getting started with Enterprise Library, and there's help for existing users as well (such as the breaking changes and migration information for previous versions) available on the Enterprise Library Community Site at *http://www.codeplex.com/entlib/*.
- For more details about the Enterprise Library Application Blocks, see the *Enterprise Library Reference Documentation* and the *Enterprise Library 6 Class Library*.
- You can download and work through the Hands-On Labs for Enterprise Library, which are available at *http://aka.ms/el6hols*.
- You can download the Enterprise Library binaries, source code, Reference Implementation, or Quick-Starts from the Microsoft Download Center at *http://go.microsoft.com/fwlink/p/?LinkID=290898*.
- To help you understand how you can use Enterprise Library application blocks, we provide a series of simple example applications that you can run and examine. To obtain the example applications, go to *http://go.microsoft.com/fwlink/p/?LinkID=304210*.

7 Banishing Validation Complication

Using the Validation Application Block

Introduction

If you happen to live in the U.S. and I told you that the original release date of version 2.0 of Enterprise Library was 13/01/2006, you'd wonder if I'd invented some new kind of calendar. Perhaps they added a new month to the calendar without you noticing (which I'd like to call Plutember in honor of the now-downgraded ninth planet). Of course, in many other countries around the world, 13/01/2006 is a perfectly valid date in January. This validation issue is well known and the solution probably seems obvious, but I once worked with an application that used dates formatted in the U.S. mm/dd/yyyy pattern for the filenames of reports it generated, and defaulted to opening the report for the previous day when you started the program. Needless to say, on my machines set up to use U.K. date formats, it never did manage to find the previous day's report.

Even better, when I tried to submit a technical support question on their Web site, it asked me for the date I purchased the software. The JavaScript validation code in the Web page running on my machine checked the format of my answer (27/04/2008) and accepted it. But the server refused to believe that there are twenty seven months in a year, and blocked my submission. I had to lie and say I purchased it on May 1 instead.

The problem is that validation can be an onerous task, especially when you need to do it in so many places in your applications, and for a lot of different kinds of values. It's extremely easy to end up with repeated code scattered throughout your classes, and yet still leave holes where unexpected input can creep into your application and possibly cause havoc.

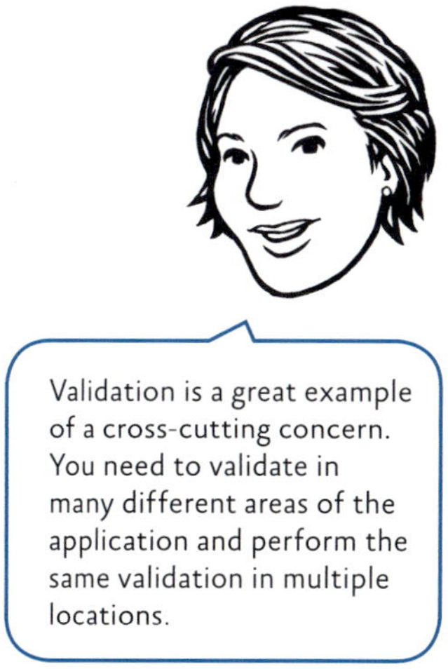

Robust validation can help to protect your application against malicious users and dangerous input (including SQL injection attacks), ensure that it processes only valid data and enforces business rules, and improve responsiveness by detecting invalid data before performing expensive processing tasks.

So, how do you implement comprehensive and centralized validation in your applications? One easy solution is to take advantage of the Enterprise Library Validation Block. The Validation block is a highly flexible solution that allows you to specify validation rules in configuration, with attributes, or in code, and have that validation applied to objects, method parameters, fields, and properties. It even includes features that integrate with Windows Forms, Windows Presentation Foundation (WPF), ASP.NET, and Windows Communication Foundation (WCF) applications to present validation errors within the user interface or have them handled automatically by the service.

Techniques for Validation

Before we explore the Validation block, it's worth briefly reviewing some validation good practices. In general, there are three factors you should consider: where you are going to perform validation, what data should you validate, and how you will perform this validation.

Where Should I Validate?

Use trust and layer boundaries in your application to identify where to add validation.

Validation should, of course, protect your entire application. However, it is often the case that you need to apply validation in more than one location. If your application consists of layers, distributed services, or discrete components, you probably need to validate at each boundary. This is especially the case where individual parts of the application could be called from more than one place (for example, a business layer that is used by several user interfaces and other services).

It is also a really good idea to validate at trust boundaries, even if the components on each side of the boundary are not physically separated. For example, your business layer may run under a different trust level or account context than your data layer (even if they reside on the same machine). Validation at this boundary can prevent code that is running in low trust and which may have been compromised, from submitting invalid data to code that runs in higher trust mode.

Finally, a common scenario: validation in the user interface. Validating data on the client can improve application responsiveness, especially if the UI is remote from the server. Users do not have to wait for the server to respond when they enter or submit invalid data, and the server does not need to attempt to process data that it will later reject. However, remember that even if you do validate data on the client or in the UI you must always revalidate on the server or in the receiving service. This protects against malicious users who may circumvent client-side validation and submit invalid data.

Validation in the UI illustrates how you can use validation to make your application more efficient. There is no point in submitting a request that will travel through several layers (possibly crossing a network) if you can identify what's wrong with the request in advance.

What Should I Validate?

Validate data from other systems as well as user input.

To put it simply, everything. Or, at least any input values you will use in your application that may cause an error, involve a security risk, or could result in incorrect processing. Remember that Web page and service requests may contain data that the user did not enter directly, but could be used in your application. This can include cookies, header information, credentials, and context information that the server may use in various ways. Treat all input data as suspicious until you have validated it.

How Should I Validate?

For maximum security, your validation process should be designed to accept only data that you can directly determine to be valid. This approach is known as positive validation and generally uses an allow list that specifies data that satisfies defined criteria, and rejects all other data. Examples are rules that check if a number is between two predefined limits, or if the submitted value is within a list of valid values. Use this approach whenever possible.

The alternative and less-secure approach is to use a block list containing values that are not valid. This is called negative validation, and generally involves accepting only data that does not meet specific criteria. For example, as long as a string does not contain any of the specified invalid characters, it would be accepted. You should use this approach cautiously and as a secondary line of defense, because it is very difficult to create a complete list of criteria for all known invalid input—which may allow malicious data to enter your system.

Finally, consider sanitizing data. While this is not strictly a validation task, you can as an extra precaution attempt to eliminate or translate characters in an effort to make the input safe. However, do not rely on this technique alone because, as with negative validation, it can be difficult to create a complete list of criteria for all known invalid input unless there is a limited range of invalid values.

What Does the Validation Block Do?

The Validation block consists of a broad range of validators, plus a mechanism that executes these validators and collects and correlates the results to provide an overall validation result (true/valid or false/invalid). The Validation block can use individual attributes applied to classes and class members that the application uses (both the validation attributes provided with the Validation block and data annotation attributes from the **System.ComponentModel.DataAnnotations** namespace), in addition to rule sets defined in the configuration of the block, which specify the validation rules to apply.

The typical scenario when using the Validation block is to define rule sets through configuration or attributes applied to your classes. Each rule set specifies the set of individual validators and combinations of these validators that implement the validation rules you wish to apply to that class. Then you use a **ValidationFactory** (or one of the equivalent implementations of this factory) to create a type validator for the class, optionally specifying the rule set it should use. If you don't specify a rule set, it uses the default rules. Then you can call the **Validate** method of the type validator. This method returns an instance of the **ValidationResults** class that contains details of all the validation errors detected. Figure 1 illustrates this process.

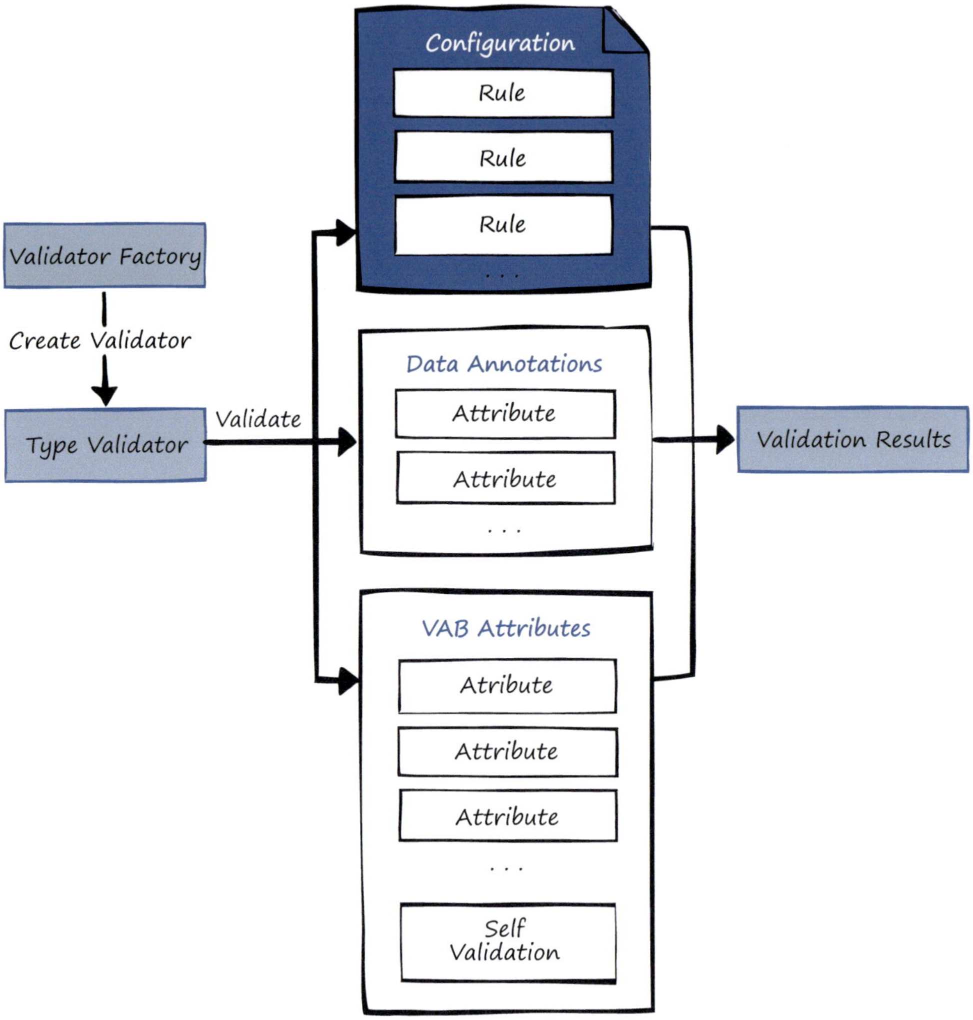

FIGURE 1
An overview of the validation process

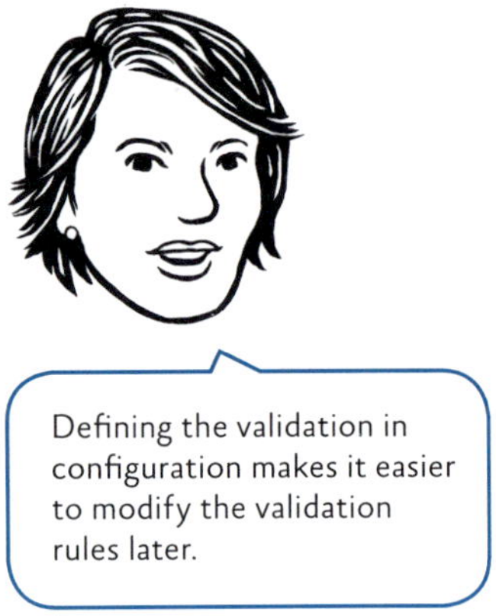

When you use a rule set to validate an instance of a specific type or object, the block can apply the rules to:

- The type itself
- The values of public readable properties
- The values of public fields
- The return values of public methods that take no parameters

Notice that you can validate the values of method parameters and the return type of methods that take parameters when that method is invoked, only by using the validation call handler (which is part of the Policy Injection Application Block) in conjunction with the Unity dependency injection and interception mechanism. The validation call handler will validate the parameter values based on the rules for each parameter type and any validation attributes applied to the parameters. We don't cover the use of the validation call handler in this guide, as it requires you to be familiar with Unity interception techniques. For more information about interception and the validation call handler, see Chapter 5, "*Interception using Unity*" in the Developer's Guide to Dependency Injection Using Unity.

Alternatively, you can create individual validators programmatically to validate specific values, such as strings or numeric values. However, this is not the main focus of the block—though we do include samples in this chapter that show how you can use individual validators.

In addition, the Validation block contains features that integrate with Windows Forms, Windows Presentation Foundation (WPF), ASP.NET, and Windows Communication Foundation (WCF) applications. These features use a range of different techniques to connect to the UI, such as a proxy validator class based on the standard ASP.NET **Validator** control that you can add to a Web page, a **ValidationProvider** class that you can specify in the properties of Windows Forms controls, a **ValidatorRule** class that you can specify in the definition of WPF controls, and a behavior extension that you can specify in the **<system.ServiceModel>** section of your WCF configuration. You'll see more details of these features later in this chapter.

The Range of Validators

Validators implement functionality for validating Microsoft .NET Framework data types. The validators included with the Validation block fall into three broad categories: value validators, composite validators, and type (object) validators. The value validators allow you to perform specific validation tests such as verifying:

- The length of a string, or the occurrence of a specified set of characters within it.
- Whether a value lies within a specified range, including tests for dates and times relative to a specified date/time.
- Whether a value is one of a specified set of values, or can be converted to a specific data type or enumeration value.
- Whether a value is null, or is the same as the value of a specific property of an object.
- Whether the value matches a specified regular expression.

The composite validators are used to combine other validators when you need to apply more complex validation rules. The Validation block includes the composite **AND** and **OR** validators, each of which acts as a container for other validators. By nesting these composite validators in any combination and populating them with other validators, you can create very comprehensive and very specific validation rules.

Table 1 describes the complete set of validators provided with the Validation block.

Validator type	Validator name	Description
Value Validators	Contains Characters Validator	Checks that an arbitrary string, such as a string entered by a user in a Web form, contains any or all of the specified characters.
	Date Time Range Validator	Checks that a DateTime object falls within a specified range.
	Domain Validator	Checks that a value is one of the specified values in a specified set.
	Enum Conversion Validator	Checks that a string can be converted to a value in a specified enumeration type.
	Not Null Validator	Checks that the value is not null.
	Property Comparison Validator	Compares the value to be checked with the value of a specified property.
	Range Validator	Checks that a value falls within a specified range.
	Regular Expression Validator	Checks that the value matches the pattern specified by a regular expression.
	Relative Date Time Validator	Checks that the DateTime value falls within a specified range using relative times and dates.
	String Length Validator	Checks that the length of the string is within the specified range.
	Type Conversion Validator	Checks that a string can be converted to a specific type.
Type Validators	Object Validator	Causes validation to occur on an object reference. All validators defined for the object's type will be invoked.
	Object Collection Validator	Checks that the object is a collection of the specified type and then invokes validation on each element of the collection.
Composite Validators	And Composite Validator	Requires all validators that make up the composite validator to be true.
	Or Composite Validator	Requires at least one of the validators that make up the composite validator be true.
Single Member Validators	Field Value Validator	Validates a field of a type.
	Method Return Value Validator	Validates the return value of a method of a type.
	Property Value Validator	Validates the value of a property of a type.

Table 1 - The validators provided with the Validation block

For more details on each validator, see the *Enterprise Library Reference Documentation*. You will see examples that use many of these validators throughout this chapter.

Validating with Attributes

If you have full access to the source code of your application, you can use attributes within your classes to define your validation rules. You can apply validation attributes in the following ways:

- To a **field**. The Validation block will check that the field value satisfies all validation rules defined in validators applied to the field.
- To a **property**. The Validation block will check that the value of the get property satisfies all validation rules defined in validators applied to the property.
- To a **method that takes no parameters**. The Validation block will check that the return value of the method satisfies all validation rules defined in validators applied to the method.

- To an entire **class**, using only the **NotNullValidator**, **ObjectCollectionValidator**, **AndCompositeValidator**, and **OrCompositeValidator**). The Validation block can check if the object is null, that it is a member of the specified collection, and that any validation rules defined within it are satisfied.
- To a **parameter in a WCF Service Contract**. The Validation block will check that the parameter value satisfies all validation rules defined in validators applied to the parameter.

To **parameters of methods that are intercepted**, by using the validation call handler in conjunction with the Policy Injection Application Block. For more information on using interception, see Chapter 5, "*Interception using Unity*" in the Unity Developer's Guide.

Each of the validators described in the previous section has a related attribute that you apply in your code, specifying the values for validation (such as the range or comparison value) as parameters to the attribute. For example, you can validate a property that must have a value between 0 and 10 inclusive by applying the following attribute to the property definition, as seen in the following code.

```
[RangeValidator(0, RangeBoundaryType.Inclusive, 10, RangeBoundaryType.Inclusive)]
```

DataAnnotations Attributes

In addition to using the built-in validation attributes, the Validation block will perform validation defined in the vast majority of the validation attributes in the **System.ComponentModel.DataAnnotations** namespace. These attributes are typically used by frameworks and object/relational mapping (O/RM) solutions that auto-generate classes that represent data items. They are also generated by the ASP.NET validation controls that perform both client-side and server-side validation. While the set of validation attributes provided by the Validation block does not map exactly to those in the **DataAnnotations** namespace, the most common types of validation are supported. A typical use of data annotations is shown here.

```
[System.ComponentModel.DataAnnotations.Required(
        ErrorMessage = "You must specify a value for the product ID.")]
[System.ComponentModel.DataAnnotations.StringLength(6,
        ErrorMessage = "Product ID must be 6 characters.")]
[System.ComponentModel.DataAnnotations.RegularExpression("[A-Z]{2}[0-9]{4}",
        ErrorMessage = "Product ID must be 2 capital letters and 4 numbers.")]
public string ID { get; set; }
```

In reality, the Validation block validation attributes are data annotation attributes, and can be used (with some limitations) whenever you can use data annotations attributes—for example, with ASP.NET Dynamic Data applications. The main difference is that the Validation block attribute validation occurs only on the server, and not on the client.

Also keep in mind that, while DataAnnotations supports most of the Validation block attributes, not all of the validation attributes provided with the Validation block are supported by the built-in .NET validation mechanism. For more information, see the *Enterprise Library Reference Documentation*, and the topic "*System.ComponentModel.DataAnnotations Namespace*" on MSDN.

Self-Validation

Self-validation might sound as though you should be congratulating yourself on your attractiveness and wisdom, and your status as fine and upstanding citizen. However, in Enterprise Library terms, self-validation is concerned with the use of classes that contain their own validation logic.

For example, a class that stores spare parts for aircraft might contain a function that checks if the part ID matches a specific format containing letters and numbers. You add the **HasSelfValidation** attribute to the class, add the **SelfValidation** attribute to any validation functions it contains, and optionally add attributes for the built-in Validation block validators to any relevant properties. Then you can validate an instance of the class using the Validation block. The block will execute the self-validation method.

Self-validation cannot be used with the UI validation integration features for Windows Forms, WPF, or ASP.NET.

Self-validation is typically used where the validation rule you want to apply involves values from different parts of your class or values that are not publicly exposed by the class, or when the validation scenario requires complex rules that even a combination of composed validators cannot achieve. For example, you may want to check if the sum of the number of products on order and the number already in stock is less than a certain value before allowing a user to order more. The following extract from one of the examples you'll see later in this chapter shows how self-validation can be used in this case.

```
[HasSelfValidation]
public class AnnotatedProduct : Iproduct
  ...
  // code to implement constructor and properties goes here
  ...

  [SelfValidation]
  public void Validate(ValidationResults results)
  {
    string msg = string.Empty;
    if (InStock + OnOrder > 100)
    {
      msg = "Total inventory (in stock and on order) cannot exceed 100 items.";
      results.AddResult(new ValidationResult(msg, this, "ProductSelfValidation", "", null));
  }
}
```

The Validation block calls the self-validation method when you validate this class instance, passing to it a reference to the collection of **ValidationResults** that it is populating with any validation errors found. The code above simply adds one or more new **ValidationResult** instances to the collection if the self-validation method detects an invalid condition. The parameters of the **ValidationResult** constructor are:

- The validation error message to display to the user or write to a log. The **ValidationResult** class exposes this as the **Message** property.
- A reference to the class instance where the validation error was discovered (usually the current instance). The **ValidationResult** class exposes this as the **Target** property.
- A string value that describes the location of the error (usually the name of the class member, or some other value that helps locate the error). The **ValidationResult** class exposes this as the **Key** property.
- An optional string tag value that can be used to categorize or filter the results. The **ValidationResult** class exposes this as the **Tag** property.
- A reference to the validator that performed the validation. This is not used in self-validation, though it will be populated by other validators that validate individual members of the type. The **ValidationResult** class exposes this as the **Validator** property.

Validation Rule Sets

A validation rule set is a combination of all the rules with the same name, which may be located in a configuration file or other configuration source, in attributes defined within the target type, and implemented through self-validation. In other words, a rule set includes any type of validation rule that has a specified name.

Rule set names are case-sensitive. *The two rule sets named* ***MyRuleset*** *and* ***MyRuleSet*** *are different!*

How do you apply a name to a validation rule? And what happens if you don't specify a name? In fact, the way it works is relatively simple, even though it may appear complicated when you look at configuration and attributes, and take into account how these are actually processed.

To start with, every validation rule is a member of some rule set. If you do not specify a name, that rule is a member of the default rule set; effectively, this is the rule set whose name is an empty string. When you do specify a name for a rule, it becomes part of the rule set with that name.

Assigning Validation Rules to Rule Sets

You specify rule set names in a variety of ways, depending on the location and type of the rule:

- **In configuration**. You define a type that you want to apply rules to, and then define one or more rule sets for that type. To each rule set you add the required combination of validators, each one representing a validation rule within that rule set. You can specify one rule set for each type as the default rule set for that type. The rules within this rule set are then treated as members of the default (unnamed) rule set, as well as that named rule set.
- **In Validation block validator attributes applied to classes and their members**. Every validation attribute will accept a rule set name as a parameter. For example, you specify that a **NotNullValidator** is a member of a rule set named **MyRuleset**, like this.

```
[NotNullValidator(MessageTemplate = "Cannot be null",
                  Ruleset = "MyRuleset")]
```

- **In SelfValidation attributes within a class**. You add the **Ruleset** parameter to the attribute to indicate which rule set this self-validation rule belongs to. You can define multiple self-validation methods in a class, and add them to different rule sets if required.

```
[SelfValidation(Ruleset = "MyRuleset")]
```

Configuring Validation Block Rule Sets

The Enterprise Library configuration console makes it easy to define rule sets for specific types that you will validate. Each rule set specifies a type to which you will apply the rule set, and allows you to specify a set of validation rules. You can then apply these rules as a complete set to an instance of an object of the defined type.

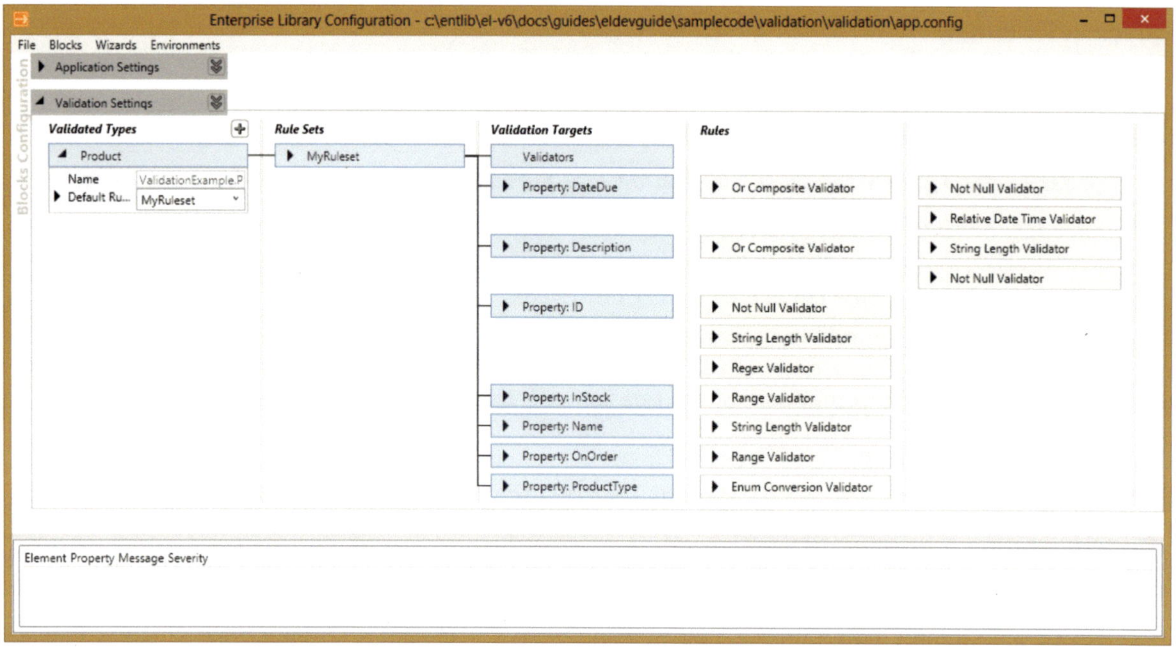

FIGURE 2
The configuration for the examples

Figure 2 shows the configuration console with the configuration used in the example application for this chapter. It defines a rule set named **MyRuleset** for the validated type (the **Product** class). **MyRuleset** is configured as the default rule set, and contains a series of validators for all of the properties of the **Product** type. These validators include two Or Composite Validators (which contain other validators) for the **DateDue** and **Description** properties, three validators that will be combined with the **And** operation for the **ID** property, and individual validators for the remaining properties.

When you highlight a rule, member, or validator in the configuration console, it shows connection lines between the configured items to help you see the relationships between them.

Specifying Rule Sets When Validating

You can specify a rule set name when you create a type validator that will validate an instance of a type. If you use the **ValidationFactory** facade to create a type validator for a type, you can specify a rule set name as a parameter of the **CreateValidator** method. If you create an Object Validator or an Object Collection Validator programmatically by calling the constructor, you can specify a rule set name as a parameter of the constructor. We look in more detail at the options for creating validators later in this chapter.

- **If you specify a rule set name** when you create a validator for an object, the Validation block will apply only those validation rules that are part of the specified rule set. It will, by default, apply all rules with the specified name that it can find in configuration, attributes, and self-validation.
- **If you do not specify a rule set name** when you create a validator for an object, the Validation block will, by default, apply all rules that have no name (effectively, rules with an empty string as the name) that it can find in configuration, attributes, and self-validation. If you have specified one rule set in configuration as the default rule set for the type you are validating (by setting the **DefaultRule** property for that type to the rule set name), rules within this rule set are also treated as being members of the default (unnamed) rule set.

The one time that this default mechanism changes is if you create a validator for a type using a facade other than **ValidationFactory**. As you'll see later in this chapter you can use the **ConfigurationValidatorFactory**, **AttributeValidator-Factory**, or **ValidationAttributeValidatorFactory** to generate type validators. In this case, the validator will only apply rules that have the specified name and exist in the specified location.

For example, when you use a **ConfigurationValidatorFactory** and specify the name **MyRuleset** as the rule set name when you call the **CreateValidator** method, the validator you obtain will only process rules it finds in configuration that are defined within a rule set named **MyRuleset** for the target object type. If you use an **AttributeValidatorFactory**, the validator will only apply Validation block rules located in attributes and self-validation methods of the target class that have the name **MyRuleset**.

Configuring multiple rule sets for the same type is useful when the type you need to validate is a primitive type such as a String. A single application may have dozens of different rule sets that all target String.

How Do I Use The Validation Block?

In the remainder of this chapter, we'll show you in more detail how you can use the features of the Validation block you have seen in previous sections. In this section, we cover three topics that you should be familiar with when you start to use the block in your applications: preparing your application to use the block, choosing a suitable approach for validation, the options available for creating validators, accessing and displaying validation errors, and understanding how you can use template tokens in validation messages.

Preparing Your Application

To use the Validation block, you must reference the required assemblies: you can use the NuGet Package Manager to add the required assemblies and references. In the **Manage Nuget Packages** dialog in Visual Studio, search online for the **EnterpriseLibrary.Validation** package and install it.

If you intend to use the integration features for ASP.NET, Windows Forms, WPF, or WCF, you must also install the relevant NuGet package that contains these features: **EnterpriseLibrary.Validation.Integration.AspNet**, **Enterprise-Library.Validation.Integration.WinForms**, **EnterpriseLibrary.Validation.Integration.WPF**,and **EnterpriseLibrary.Validation.Integration.WCF**.

Then you can edit your code to specify the namespaces used by the Validation block and, optionally, the integration features if you need to integrate with WCF or a UI technology.

> *If you are using WCF integration, you should also add a reference to the System.ServiceModel namespace.*

Choosing a Validation Approach

Before you start to use the Validation block, you should consider how you want to perform validation. As you've seen, there are several approaches you can follow. Table 2 summarizes these, and will help you to choose one, or a combination, most suited to your requirements.

Validation approach	Advantages	Considerations
Rule sets in configuration	Supports the full capabilities of the Validation block validators. Validation rules can be changed without requiring recompilation and redeployment. Validation rules are more visible and easier to manage.	Rules are visible in configuration files unless the content is encrypted. May be open to unauthorized alteration if not properly protected. Type definitions and validation rule definitions are stored in different files and accessed using different tools, which can be confusing.
Validation block attributes	Supports the full capabilities of the Validation block validators. Validation attributes may be defined in separate metadata classes. Rules can be extracted from the metadata for a type by using reflection.	Requires modification of the source code of the types to validate. Some complex rule combinations may not be possible—only a single And or Or combination is available for multiple rules. Hides validation rules from administrators and operators.
Data annotation attributes	Allows you to apply validation rules defined by .NET data annotation attributes, which may be defined in separate metadata classes. Typically used with technologies such as LINQ, for which supporting tools might be used to generate code. Technologies such as ASP.NET Dynamic Data that use these attributes can perform partial client-side validation. Rules can be extracted from the metadata for a type by using reflection.	Requires modification of the source code. Does not support all of the powerful validation capabilities of the Validation block.
Self-validation	Allows you to create custom validation rules that may combine values of different members of the class.	Requires modification of the source code. Hides validation rules from administrators and operators. Rules cannot be extracted from the metadata for a type by using reflection.
Validators created programmatically	A simple way to validate individual values as well as entire objects. Useful if you only need to perform small and specific validation tasks, especially on value types held in variables.	Requires additional code and is generally more difficult to manage for complex validation scenarios. Hides validation rules from administrators and operators. More difficult to administer and manage.

Table 2 - Validation approaches

If you decide to use attributes to define your validation rules within classes but are finding it difficult to choose between using the Validation block attributes and the Microsoft .NET data annotation attributes, you should consider using the Validation block attributes approach as this provides more powerful capabilities and supports a far wider range of validation operations. However, you should consider the data annotations approach in the following scenarios:

Remember, you can use a combination of approaches if you want, but plan it carefully. If you choose to create validators programmatically, you could expose selected properties through a configuration file if you wrote custom code to load values from the configuration file.

- When you are working with existing applications that already use data annotations.
- When you require validation to take place on the client.
- When you are building a Web application where you will use the ASP.NET Data Annotation Model Binder, or you are using ASP.NET Dynamic Data to create data-driven user interfaces.
- When you are using a framework such as the Microsoft Entity Framework, or another object/relational mapping (O/RM) technology that auto-generates classes that include data annotations.

Options for Creating Validators Programmatically

There are several ways that you can create the validators you require, whether you are creating a type validator that will validate an instance of your class using a rule set or attributes, or you are creating individual value validators:

- **Use the ValidationFactory facade to create validators**. This approach makes it easy to create type validators that you can use, in conjunction with rule sets, to validate multiple members of an object instance. This is generally the recommended approach. You also use this approach to create validators that use only validation attributes or data annotations within the classes you want to validate, or only rule sets defined in configuration.
- **Create individual validators programmatically** by calling their constructor. The constructor parameters allow you to specify most of the properties you require for the validator. You can then set additional properties, such as the **Tag** or the resource name and type if you want to use a resource file to provide the message template. You can also build combinations of validators using this approach to implement complex validation rules.

Performing Validation and Displaying Validation Errors

To initiate validation, you call the **Validate** method of your validator. There are two overloads of this method: one that creates and returns a populated **ValidationResults** instance, and one that accepts an existing **ValidationResults** instance as a parameter. The second overload allows you to perform several validation operations, and collect all of the errors in one **ValidationResults** instance.

You can check if validation succeeded, or if any validation errors were detected, by examining the **IsValid** property of a **ValidationResults** instance, and displaying details of any validation errors that occurred. The following code shows a simple example of how you can display the most relevant details of each validation error. See the section on self-validation earlier in this chapter for a description of the properties of each individual **ValidationResult** within the **ValidationResults**.

```
// Check if the ValidationResults detected any validation errors.
if (results.IsValid)
{
  Console.WriteLine("There were no validation errors.");
}
else
{
  Console.WriteLine("The following {0} validation errors were detected:",
                    results.Count);
  // Iterate through the collection of validation results.
  foreach (ValidationResult item in results)
  {
    // Show the target member name and current value.
    Console.WriteLine("Target:'{0}' Key:'{1}' Tag:'{2}' Message:'{3}'",
                      item.Target, item.Key, item.Tag, item.Message);
  }
}
```

Alternatively, you can extract more information about the validation result for each individual validator where an error occurred. The example application we provide demonstrates how you can do this, and you'll see more details later in this chapter.

Understanding Message Template Tokens

One specific and very useful feature of the individual validators you define in your configuration or attributes is the capability to include tokens in the message to automatically insert values of the validator's properties. This applies no matter how you create your validator—in rule sets defined in configuration, as validation attributes, or when you create validators programmatically.

The **Message** property of a validator is actually a template, not just a simple text string that is displayable. When the block adds an individual **ValidationResult** to the **ValidationResults** instance for each validation error it detects, it parses the value of the **Message** property looking for tokens that it will replace with the value of specific properties of the validator that detected the error.

> If you display or use the **Target** token, be sure to escape it to guard against injection attacks.

The value injected into the placeholder tokens, and the number of tokens used, depends on the type of validator—although there are three tokens that are common to all validators. The token **{0}** will be replaced by the value of the object being validated (ensure that you escape this value before you display or use it in order to guard against injection attacks). The token **{1}** will contain the name of the member that was being validated, if available, and is equivalent to the **Key** property of the validator. The token **{2)** will contain the value of the **Tag** property of the validator.

The remaining tokens depend the on the individual validator type. For example, in the case of the Contains Characters validator, the tokens **{3}** and **{4}** will contain the characters to check for and the **ContainsCharacters** value (**All** or **Any**). In the case of a range validator, such as the String Length validator, the tokens **{3}** to **{6}** will contain the values and bound types (**Inclusive**, **Exclusive**, or **Ignore**) for the lower and upper bounds you specify for the validator. For example, you may define a String Length validator like this:

```
[StringLengthValidator(5, RangeBoundaryType.Inclusive, 20,
        RangeBoundaryType.Inclusive,
        MessageTemplate = "{1} must be between {3} and {5} characters.")]
```

If this validator is attached to a property named **Description**, and the value of this property is invalid, the **ValidationResults** instance will contain the error message **Description must be between 5 and 20 characters**.

Other validators use tokens that are appropriate for the type of validation they perform. The documentation installed with Enterprise Library lists the tokens for each of the Validation block validators. You will also see the range of tokens used in the examples that follow.

Diving in With Some Simple Examples

The remainder of this chapter shows how you can use the Validation block in a variety of situations within your applications. We provide a simple console-based example application implemented as the Microsoft Visual Studio solution named **Validation**. You can open it in Visual Studio to view the code, or run the application directly from the bin\Debug folder.

The application uses three versions of a class that stores product information. All of these implement an interface named **IProduct**, as illustrated in Figure 3. Each has a string property that is designed to be set to a value from an enumeration called **ProductType** that defines the valid set of product type names.

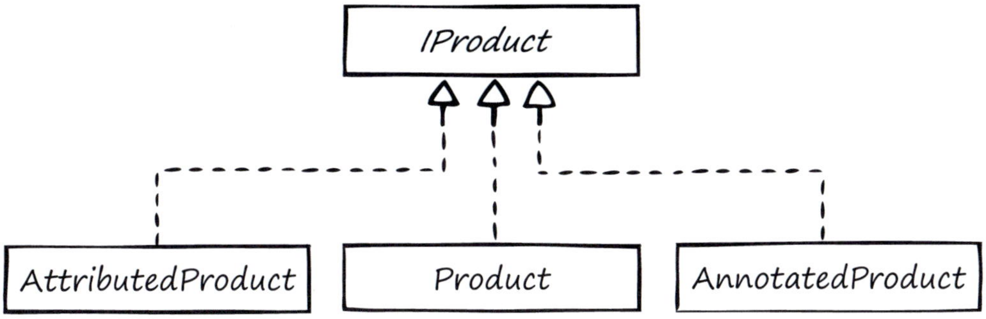

Figure 3
The product classes used in the examples

The **Product** class is used primarily with the example that demonstrates using a configured rule set, and contains no validation attributes. The **AttributedProduct** class contains Validation block attributes, while the **Annotated-Product** class contains .NET Data Annotation attributes. The latter two classes also contain self-validation routines—the extent depending on the capabilities of the type of validation attributes they contain. You'll see more on this topic when we look at the use of validation attributes later in this chapter.

The following sections of this chapter will help you understand in detail the different ways that you can use the Validation block:

- **Validating Objects and Collections of Objects**. This is the core topic for using the Validation block, and is likely to be the most common scenario in your applications. It shows how you can create type validators to validate instances of your custom classes, how you can dive deeper into the **ValidationResults** instance that is returned, how you can use the Object Validator, and how you can validate collections of objects.
- **Using Validation Attributes**. This section describes how you can use attributes applied to your classes to enable validation of members of these classes. These attributes use the Validation block validators and the .NET Data Annotation attributes.
- **Creating and Using Individual Validators**. This section shows how you can create and use the validators provided with the block to validate individual values and members of objects.
- **WCF Service Validation Integration**. This section describes how you can use the block to validate parameters within a WCF service.

Finally, we'll round off the chapter by looking briefly at how you can integrate the Validation block with user interface technologies such as Windows Forms, WPF, and ASP.NET.

Validating Objects and Collections of Objects

The most common scenario when using the Validation block is to validate an instance of a class in your application. The Validation block uses the combination of rules defined in a rule set and validators added as attributes to test the values of members of the class, and the result of executing any self-validation methods within the class.

The Validation block makes it easy to validate entire objects (all or a subset of its members) using a specific type validator or by using the Object validator. You can also validate all of the objects in a collection using the Object Collection validator. We will look at the Object validator and the Object Collection validator later. For the moment, we'll concentrate on creating and using a specific type validator.

Creating a Type Validator using the ValidationFactory

You can configure the static **ValidationFactory** type from the configuration in your .config file and use it to create a validator for a specific target type. This validator will validate objects using a rule set, and/or any attributes and self-validation methods the target object contains. To configure the **ValidationFactory** class, you can use the following code.

```
ValidationFactory.SetDefaultConfigurationValidatorFactory(
                                      new SystemConfigurationSource(false));
```

You can then create a validator for any type you want to validate. For example, this code creates a validator for the **Product** class and then validates an instance of that class named **myProduct**.

```
Validator<Product> pValidator = ValidationFactory.
                              CreateValidator<Product>();
ValidationResults valResults = pValidator.Validate(myProduct);
```

By default, the validator will use the default rule set defined for the target type (you can define multiple rule sets for a type, and specify one of these as the default for this type). If you want the validator to use a specific rule set, you specify this as the single parameter to the **CreateValidator** method, as shown here.

```
Validator<Product> productValidator
            = ValidationFactory.CreateValidator<Product>("RuleSetName");
ValidationResults valResults = productValidator.Validate(myProduct);
```

The example named *Using a Validation Rule Set to Validate an Object* creates an instance of the **Product** class that contains invalid values for all of the properties, and then uses the code shown above to create a type validator for this type and validate it. It then displays details of the validation errors contained in the returned **ValidationResults** instance. However, rather than using the simple technique of iterating over the **ValidationResults** instance displaying the top-level errors, it uses code to dive deeper into the results to show more information about each validation error, as you will see in the next section.

Delving Deeper into ValidationResults

Typically, you can get all the detail you need by iterating over the **ValidationResults** instance.

You can check if validation succeeded, or if any validation error were detected, by examining the **IsValid** property of a **ValidationResults** instance and displaying details of any validation errors that occurred. However, when you simply iterate over a **ValidationResults** instance (as we demonstrated in the section "Performing Validation and Displaying Validation Errors" earlier in this chapter), we displayed just the top-level errors. In many cases, this is all you will require. If the validation error occurs due to a validation failure in a composite (**And** or **Or**) validator, the error this approach will display is the message and details of the composite validator.

However, sometimes you may wish to delve deeper into the contents of a **ValidationResults** instance to learn more about the errors that occurred. This is especially the case when you use nested validators inside a composite validator. The code we use in the example provides richer information about the errors. When you run the example, it displays the following results (we've removed some repeated content for clarity).

```
The following 6 validation errors were detected:
+ Target object: Product, Member: DateDue
  - Detected by: OrCompositeValidator
  - Tag value: Date Due
  - Validated value was: '23/11/2010 13:45:41'
  - Message: 'Date Due must be between today and six months time.'
  + Nested validators:
    - Detected by: NotNullValidator
    - Validated value was: '23/11/2010 13:45:41'
    - Message: 'Value can be NULL or a date.'
    - Detected by: RelativeDateTimeValidator
    - Validated value was: '23/11/2010 13:45:41'
    - Message: 'Value can be NULL or a date.'
+ Target object: Product, Member: Description
  - Detected by: OrCompositeValidator
  - Validated value was: '-'
  - Message: 'Description can be NULL or a string value.'
  + Nested validators:
    - Detected by: StringLengthValidator
    - Validated value was: '-'
    - Message: 'Description must be between 5 and 100 characters.'
    - Detected by: NotNullValidator
    - Validated value was: '-'
    - Message: 'Value can be NULL.'
...
...
+ Target object: Product, Member: ProductType
  - Detected by: EnumConversionValidator
  - Tag value: Product Type
  - Validated value was: 'FurryThings'
  - Message: 'Product Type must be a value from the 'ProductType' enumeration.'
```

You can see that this shows the target object type and the name of the member of the target object that was being validated. It also shows the type of the validator that performed the operation, the **Tag** property values, and the validation error message. Notice also that the output includes the validation results from the validators nested within the two **OrCompositeValidator** validators. To achieve this, you must iterate recursively through the **ValidationResults** instance because it contains nested entries for the composite validators.

The code we used also contains a somewhat contrived feature: to be able to show the value being validated, some examples that use this routine include the validated value at the start of the message using the **{0}** token in the form: **[{0}]** ***validation error message***. The example code parses the **Message** property to extract the value and the message when it detects that this message string contains such a value. It also encodes this value for display in case it contains malicious content.

While this may not represent a requirement in real-world application scenarios, it is useful here as it allows the example to display the invalid values that caused the validation errors and help you understand how each of the validators works. We haven't listed the code here, but you can examine it in the example application to see how it works, and adapt it to meet your own requirements. You'll find it in the **ShowValidationResults**, **ShowValidatorDetails**, and **GetTypeNameOnly** routines located in the region named Auxiliary routines at the end of the main program file.

Using the Object Validator

An alternative approach to validating objects is to programmatically create an Object Validator by calling its constructor. You specify the type that it will validate and, optionally, a rule set to use when performing validation. If you do not specify a rule set name, the validator will use the default rule set. When you call the **Validate** method of the Object validator, it creates a type-specific validator for the target type you specify, and you can use this to validate the object, as shown here.

```
Validator pValidator = new ObjectValidator(typeof(Product), "RuleSetName");
ValidationResults valResults = pValidator.Validate(myProduct);
```

Alternatively, you can call the default constructor of the Object validator. In this case, it will create a type-specific validator for the type of the target instance you pass to the **Validate** method. If you do not specify a rule set name in the constructor, the validation will use the default rule set defined for the type it is validating.

```
Validator pValidator = new ObjectValidator(ValidationFactory.DefaultCompositeValidatorFactory, "RuleSetName");
ValidationResults valResults = pValidator.Validate(myProduct);
```

The validation will take into account any applicable rule sets, and any attributes and self-validation methods found within the target object.

Differences Between the Object Validator and the Factory-Created Type Validators

While the two approaches you've just seen to creating or obtaining a validator for an object achieve the same result, there are some differences in their behavior:

- If you do not specify a target type when you create an Object Validator programmatically, you can use it to validate any type. When you call the **Validate** method, you specify the target instance, and the Object validator creates a type-specific validator for the type of the target instance. In contrast, the validator you obtain from a factory can only be used to validate instances of the type you specify when you obtain the validator. However, it can also be used to validate subclasses of the specified type, but it will use the rules defined for the specified target type.
- The Object Validator will always use rules in configuration for the type of the target object, and attributes and self-validation methods within the target instance. In contrast, you can use a specific factory class type to obtain validators that only validate the target instance using one type of rule source (in other words, just configuration rule sets, or just one type of attributes).
- The Object Validator will acquire a type-specific validator of the appropriate type each time you call the **Validate** method, even if you use the same instance of the Object validator every time. In contrast, a validator obtained from one of the factory classes does not need to do this, and will offer improved performance.

As you can see from the flexibility and performance advantages listed above, you should generally consider using the **ValidationFactory** approach for creating validators to validate objects rather than creating individual Object Validator instances.

Validating Collections of Objects

Before we leave the topic of validation of objects, it is worth looking at how you can validate collections of objects. The Object Collection validator can be used to check that every object in a collection is of the specified type, and to perform validation on every member of the collection. You can apply the Object Collection validator to a property of a class that is a collection of objects using a Validation block attribute if you wish, as shown in this example that ensures that the **ProductList** property is a collection of **Product** instances, and that every instance in the collection contains valid values.

```
[ObjectCollectionValidator(typeof(Product))]
public Product[] ProductList { get; }
```

You can also create an Object Collection validator programmatically, and use it to validate a collection held in a variable. The example named *Validating a Collection of Objects* demonstrates this approach. It creates a **List** named **productList** that contains two instances of the **Product** class, one of which contains all valid values, and one that contains invalid values for some of its properties. Next, the code creates an Object Collection validator for the **Product** type and then calls the **Validate** method.

```
// Create an Object Collection Validator for the collection type.
Validator collValidator
          = new ObjectCollectionValidator(typeof(Product));

// Validate all of the objects in the collection.
ValidationResults results = collValidator.Validate(productList);
```

Finally, the code displays the validation errors using the same routine as in earlier examples. As the invalid Product instance contains the same values as the previous example, the result is the same. You can run the example and view the code to verify that this is the case.

Using Validation Attributes

Having seen how you can use rule sets defined in configuration, and how you can display the results of a validation process, we can move on to explore the other ways you can define validation rules in your applications. The example application contains two classes that contain validation attributes and a self-validation method. The **AttributedProduct** class contains Validation block attributes, while the **AnnotatedProduct** class contains data annotation attributes.

Using the Validation Block Attributes

The topic, "*Using Self Validation*" in the Enterprise Library Reference Documentation, demonstrates use of the Validation block attributes. The **AttributedProduct** class has a range of different Validation block attributes applied to the properties of the class, applying the same rules as the **MyRuleset** rule set defined in configuration and used in the previous examples.

For example, the **ID** property carries attributes that add a Not Null validator, a String Length validator, and a Regular Expression validator. These validation rules are, by default, combined with an **And** operation, so all of the conditions must be satisfied if validation will succeed for the value of this property.

```
[NotNullValidator(MessageTemplate = "You must specify a product ID.")]
[StringLengthValidator(6, RangeBoundaryType.Inclusive,
                       6, RangeBoundaryType.Inclusive,
                       MessageTemplate = "Product ID must be {3} characters.")]
[RegexValidator("[A-Z]{2}[0-9]{4}",
                MessageTemplate = "Product ID must be 2 letters and 4 numbers.")]
public string ID { get; set; }
```

Other validation attributes used within the **AttributedProduct** class include an Enum Conversion validator that ensures that the value of the **ProductType** property is a member of the **ProductType** enumeration, shown here. Note that the token **{3}** for the String Length validator used in the previous section of code is the lower bound value, while the token **{3}** for the Enum Conversion validator is the name of the enumeration it is comparing the validated value against.

```
[EnumConversionValidator(typeof(ProductType),
MessageTemplate = "Product type must be a value from the '{3}' enumeration.")]

public string ProductType { get; set; }
```

Combining Validation Attribute Operations

One other use of validation attributes worth a mention here is the application of a composite validator. By default, multiple validators defined for a member are combined using the **And** operation. If you want to combine multiple validation attributes using an Or operation, you must apply the **ValidatorComposition** attribute first and specify **CompositionType.Or**. The results of all validation operations defined in subsequent validation attributes are combined using the operation you specify for composition type.

The example class uses a **ValidatorComposition** attribute on the nullable **DateDue** property to combine a Not Null validator and a Relative DateTime validator. The top-level error message that the user will see for this property (when you do not recursively iterate through the contents of the **ValidationResults**) is the message from the **ValidatorComposition** attribute.

Using the validation attributes, you can either use the default **And** operation for all the validators on a member or explicitly use the **Or** operation for all the validators. You cannot have complex nested validators as you can when you define the validation rules in your configuration file.

```
[ValidatorComposition(CompositionType.Or,
        MessageTemplate = "Date due must be between today and six months
                           time.")]
[NotNullValidator(Negated = true,
        MessageTemplate = "Value can be NULL or a date.")]
[RelativeDateTimeValidator(0, DateTimeUnit.Day, 6, DateTimeUnit.Month,
        MessageTemplate = "Value can be NULL or a date.")]
public DateTime? DateDue { get; set; }
```

If you want to allow null values for a member of a class, you can apply the **IgnoreNulls** attribute.

Applying Self-Validation

Some validation rules are too complex to apply using the validators provided with the Validation block or the .NET Data Annotation validation attributes. It may be that the values you need to perform validation come from different places, such as properties, fields, and internal variables, or involve complex calculations.

Use self-validation rules to define complex rules that are impossible or difficult to define with the standard validators.

In this case, you can define self-validation rules as methods within your class (the method names are irrelevant), as described earlier in this chapter in the section "Self-Validation." We've implemented a self-validation routine in the **AttributedProduct** class in the example application. The method simply checks that the combination of the values of the **InStock**, **OnOrder**, and **DateDue** properties meets predefined rules. You can examine the code within the **AttributedProduct** class to see the implementation.

Results of the Validation Operation

The example creates an invalid instance of the **AttributedProduct** class shown above, validates it, and then displays the results of the validation process. It creates the following output, though we have removed some of the repeated output here for clarity. You can run the example yourself to see the full results.

```
Created and populated a valid instance of the AttributedProduct class.
There were no validation errors.

Created and populated an invalid instance of the AttributedProduct class.
The following 7 validation errors were detected:
+ Target object: AttributedProduct, Member: ID
  - Detected by: RegexValidator
  - Validated value was: '12075'
  - Message: 'Product ID must be 2 capital letters and 4 numbers.'
...
...
+ Target object: AttributedProduct, Member: ProductType
  - Detected by: EnumConversionValidator
  - Validated value was: 'FurryThings'
  - Message: 'Product type must be a value from the 'ProductType'
             enumeration.'
...
...
```

```
+ Target object: AttributedProduct, Member: DateDue
  - Detected by: OrCompositeValidator
  - Validated value was: '19/08/2010 15:55:16'
  - Message: 'Date due must be between today and six months time.'
  + Nested validators:
    - Detected by: RelativeDateTimeValidator
    - Validated value was: '18/11/2010 13:36:02'
    - Message: 'Value can be NULL or a date.'
    - Detected by: NotNullValidator
    - Validated value was: '18/11/2010 13:36:02'
+ Target object: AttributedProduct, Member: ProductSelfValidation
  - Detected by: [none]
  - Tag value:
  - Message: 'Total inventory (in stock and on order) cannot exceed 100 items.'
```

Notice that the output includes the name of the type and the name of the member (property) that was validated, as well as displaying type of validator that detected the error, the current value of the member, and the message. For the **DateDue** property, the output shows the two validators nested within the Or Composite validator. Finally, it shows the result from the self-validation method. The values you see for the self-validation are those the code in the self-validation method specifically added to the **ValidationResults** instance.

Validating Subclass Types

While discussing validation through attributes, we should briefly touch on the factors involved when you validate a class that inherits from the type you specified when creating the validator you use to validate it. For example, if you have a class named **SaleProduct** that derives from **Product**, you can use a validator defined for the **Product** class to validate instances of the **SaleProduct** class. The **Validate** method will also apply any relevant rules defined in attributes in both the **SaleProduct** class and the **Product** base class.

If the derived class inherits a member from the base class and does not override it, the validators for that member defined in the base class apply to the derived class. If the derived class inherits a member but overrides it, the validators defined in the base class for that member do not apply to the derived class.

Validating Properties that are Objects

In many cases, you may have a property of your class defined as the type of another class. For example, your **OrderLine** class is likely to have a property that is a reference to an instance of the **Product** class. It's common for this property to be defined as a base type or interface type, allowing you to set it to an instance of any class that inherits or implements the type specified for the property.

You can validate such a property using an **ObjectValidator** attribute within the class. However, by default, the validator will validate the property using rules defined for the type of the property—in this example the type **IProduct**. If you want the validation to take place based on the actual type of the object that is currently set as the value of the property, you can add the **ValidateActualType** parameter to the **ObjectValidator** attribute, as shown here.

```
public class OrderLine
{
  [ObjectValidator(ValidateActualType=true)]
  public IProduct OrderProduct { get; set; }
  ...
}
```

Using Data Annotation Attributes

The System.ComponentModel.DataAnnotations namespace in the .NET Framework contains a series of attributes that you can add to your classes and class members to signify metadata for these classes and members. They include a range of validation attributes that you can use to apply validation rules to your classes in much the same way as you can with the Validation block attributes. For example, the following shows how you can use the Range attribute to specify that the value of the property named **OnOrder** must be between 0 and 50.

```
[Range(0, 50, ErrorMessage = "Quantity on order must be between 0 and 50.")]
public int OnOrder { get; set; }
```

Compared to the validation attributes provided with the Validation block, there are some limitations when using the validation attributes from the DataAnnotations namespace:

- The range of supported validation operations is less comprehensive, though there are some new validation types available in.NET Framework 4.0 and 4.5 that extend the range. However, some validation operations such as enumeration membership checking, and relative date and time comparison are not available when using data annotation validation attributes.
- There is no capability to use **Or** composition, as there is with the Or Composite validator in the Validation block. The only composition available with data annotation validation attributes is the **And** operation.
- You cannot specify rule sets names, and so all rules implemented with data annotation validation attributes belong to the default rule set.
- There is no simple built-in support for self-validation, as there is in the Validation block.

You can, of course, include both data annotation and Validation block attributes in the same class if you wish, and implement self-validation using the Validation block mechanism in a class that contains data annotation validation attributes. The validation methods in the Validation block will process both types of attributes.

For more information about data annotations, see "*System.ComponentModel.DataAnnotations Namespace*" on MSDN.

An Example of Using Data Annotations

The examples we provide for this chapter include one named *Using Data Annotation Attributes and Self-Validation*. The class named **AnnotatedProduct** contains data annotation attributes to implement the same rules as those applied by Validation block attributes in the **AttributedProduct** class (which you saw in the previous example). However, due to the limitations with data annotations, the self-validation method within the class has to do more work to achieve the same validation rules.

The self-validation method has to check the value of the **DateDue** property to ensure it is not more than six months in the future, and that the value of the **ProductType** property is a member of the **ProductType** enumeration. To perform the enumeration check, the self-validation method creates an instance of the Validation block Enum Conversion validator programmatically, and then calls its **DoValidate** method (which allows you to pass in all of the values required to perform the validation). The code passes to this method the value of the **ProductType** property, a reference to the current object, the name of the enumeration, and a reference to the **ValidationResults** instance being use to hold all of the validation errors.

```
var enumConverterValidator = new EnumConversionValidator(typeof(ProductType),
                                "Product type must be a value from the '{3}' enumeration.");
enumConverterValidator.DoValidate(ProductType, this, "ProductType", results);
```

The code that creates the object to validate, validates it, and then displays the results, is the same as you saw in the previous example, with the exception that it creates an invalid instance of the **AnnotatedProduct** class, rather than the **AttributedProduct** class. The result when you run this example is also similar to that of the previous example, but with a few exceptions. We've listed some of the output here.

```
Created and populated an invalid instance of the AnnotatedProduct class.
The following 7 validation errors were detected:
+ Target object: AnnotatedProduct, Member: ID
  - Detected by: [none]
  - Tag value:
  - Message: 'Product ID must be 6 characters.'
...
+ Target object: AnnotatedProduct, Member: ProductSelfValidation
  - Detected by: [none]
  - Tag value:
  - Message: 'Total inventory (in stock and on order) cannot exceed 100
items.'
+ Target object: AnnotatedProduct, Member: ID
  - Detected by: ValidationAttributeValidator
  - Message: 'Product ID must be 2 capital letters and 4 numbers.'
+ Target object: AnnotatedProduct, Member: InStock
  - Detected by: ValidationAttributeValidator
  - Message: 'Quantity in stock cannot be less than 0.'
```

You can see that validation failures detected for data annotations contain less information than those detected for the Validation block attributes, and validation errors are shown as being detected by the **Validation-AttributeValidator** class—the base class for data annotation validation attributes. However, where we performed additional validation using the self-validation method, there is extra information available.

Defining Attributes in Metadata Classes

In some cases, you may want to locate your validation attributes (both Validation block attributes and .NET Data Annotation validation attributes) in a file separate from the one that defines the class that you will validate. This is a common scenario when you are using tools that generate the class files, and would therefore overwrite your validation attributes. To avoid this you can locate your validation attributes in a separate file that forms a partial class along with the main class file. This approach makes use of the **MetadataType** attribute from the System.ComponentModel.DataAnnotations namespace.

If the classes you want to validate are automatically generated, use partial classes to add the validation attributes so that they don't get overwritten.

You apply the **MetadataType** attribute to your main class file, specifying the type of the class that stores the validation attributes you want to apply to your main class members. You must define this as a partial class, as shown here. The only change to the content of this class compared to the attributed versions you saw in the previous sections of this chapter is that it contains no validation attributes.

```
[MetadataType(typeof(ProductMetadata))]
public partial class Product
{
  ... // Existing members defined here, but without attributes or annotations ...
}
```

You then define the metadata type as a normal class, except that you declare simple properties for each of the members to which you want to apply validation attributes. The actual type of these properties is not important, and is ignored by the compiler. The accepted approach is to declare them all as type **Object**. As an example, if your **Product** class contains the **ID** and **Description** properties, you can define the metadata class for it, as shown here.

```
public class ProductMetadata
{
  [Required(ErrorMessage = "ID is required.")]
  [RegularExpression("[A-Z]{2}[0-9]{4}",
                     ErrorMessage = "Product ID must be 2 capital letters and 4 numbers.")]
  public object ID;

  [StringLength(100, ErrorMessage = "Description must be less than 100 chars.")]
  public object Description;
}
```

Specifying the Location of Validation Rules

When you use a validator obtained from the **ValidationFactory**, as we've done so far in the example, validation will take into account any applicable rule sets defined in configuration and in attributes and self-validation methods found within the target object. However, you can use different factory types if you want to perform validation using only rule sets defined in configuration, or using only attributes and self-validation. The specialized types of factory you can use are:

- **ConfigurationValidatorFactory**. This factory creates validators that only apply rules defined in a configuration file, or in a configuration source you provide. By default it looks for configuration in the default configuration file (App.config or Web.config). However, you can create an instance of a class that implements the **IConfigurationSource** interface, populate it with configuration data from another file or configuration storage media, and use this when you create this validator factory.
- **AttributeValidatorFactory**. This factory creates validators that only apply rules defined in Validation block attributes located in the target class, and rules defined through self-validation methods.
- **ValidationAttributeValidatorFactory**. This factory creates validators that only apply rules defined in . NET Data Annotations validation attributes.

For example, to obtain a validator for the **Product** class that validates using only attributes and self-validation methods within the target instance, and validate an instance of this class, you create an instance of the **AttributeValidatorFactory**, as shown here.

```
AttributeValidatorFactory attrFactory = new AttributeValidatorFactory();

Validator<Product> pValidator = attrFactory.CreateValidator<Product>();
ValidationResults valResults = pValidator.Validate(myProduct);
```

Creating and Using Individual Validators

You can create an instance of any of the validators included in the Validation block directly in your code, and then call its **Validate** method to validate an object or value. For example, you can create a new Date Time Range validator and set the properties, such as the upper and lower bounds, the message, and the **Tag** property. Then you call the **Validate** method of the validator, specifying the object or value you want to validate. The example, *Creating and Using Validators Directly*, demonstrates the creation and use of some of the individual and composite validators provided with the Validation block.

Validating Strings for Contained Characters

The example code first creates a **ContainsCharactersValidator** that specifies that the validated value must contain the characters c, a, and t, and that it must contain all of these characters (you can, if you wish, specify that it must only contain **Any** of the characters). The code also sets the **Tag** property to a user-defined string that helps to identify the validator in the list of errors. The overload of the **Validate** method used here returns a new **ValidationResults** instance containing a **ValidationResult** instance for each validation error that occurred.

```
// Create a Contains Characters Validator and use it to validate a string.
Validator charsValidator = new ContainsCharactersValidator("cat",
                                ContainsCharacters.All,
                                " Value must contain {4} of the characters '{3}'.");
charsValidator.Tag = "Validating the String value 'disconnected'";
ValidationResults valResults = charsValidator.Validate("disconnected");
```

Validating Integers within a Domain

Next, the example code creates a new **DomainValidator** for integer values, specifying an error message and an array of acceptable values. Then it can be used to validate an integer, with a reference to the existing **ValidationResults** instance passed to the **Validate** method this time.

```
// Create a Domain Validator and use it to validate an Integer value.
Validator integerValidator = new DomainValidator<int>(
                                  "Value must be in the list 1, 3, 7, 11, 13.",
                                  new int[] {1, 3, 7, 11, 13});
integerValidator.Tag = "Validating the Integer value '42'";
integerValidator.Validate(42, valResults);
```

Validating with a Composite Validator

To show how you can create composite validators, the next section of the example creates an array containing two validators: a **NotNullValidator** and a **StringLengthValidator**. The first parameter of the **NotNullValidator** sets the **Negated** property. In this example, we set it to true so that the validator will allow null values. The **StringLengthValidator** specifies that the string it validates must be exactly five characters long. Notice that range validators such as the **StringLengthValidator** have properties that specify not only the upper and lower bound values, but also whether these values are included in the valid result set (**RangeBoundaryType.Inclusive**) or excluded (**RangeBoundaryType.Exclusive**). If you do not want to specify a value for the upper or lower bound of a range validator, you must set the corresponding property to **RangeBoundaryType.Ignore**.

```
Validator[] valArray = new Validator[]
{
  new NotNullValidator(true, "Value can be NULL."),
  new StringLengthValidator(5, RangeBoundaryType.Inclusive,
                            5, RangeBoundaryType.Inclusive,
                            "Must be between {3} ({4}) and {5} ({6}) chars.")
};
```

Having created an array of validators, we can now use this to create a composite validator. There are two composite validators, the **AndCompositeValidator** and the **OrCompositeValidator**. You can combine these as well to create any nested hierarchy of validators you require, with each combination returning a valid result if all (with the **AndCompositeValidator**) or any (with the **OrCompositeValidator**) of the validators it contains are valid. The example creates an **OrCompositeValidator**, which will return true (valid) if the validated string is either null or contains exactly five characters. Then it validates a null value and an invalid string, passing into the **Validate** method the existing **ValidationResults** instance.

```
Validator orValidator = new OrCompositeValidator(
                            "Value can be NULL or a string of 5 characters.",
                             valArray);

// Validate two values with the Or Composite Validator.
orValidator.Validate(null, valResults);
orValidator.Validate("MoreThan5Chars", valResults);
```

Validating Single Members of an Object

The Validation block contains three validators you can use to validate individual members of a class directly, instead of validating the entire type using attributes or rule sets. Although you may not use this approach very often, you might find it to be useful in some scenarios. The Field Value validator can be used to validate the value of a field of a type. The Method Return Value validator can be used to validate the return value of a method of a type. Finally, the Property Value validator can be used to validate the value of a property of a type.

Typically, you want to validate a complete instance, but if you want to validate just a single method or property, you can use the Method Return Value or the Property Value validators.

The example shows how you can use a Property Value validator. The code creates an instance of the **Product** class that has an invalid value for the ID property, and then creates an instance of the **PropertyValueValidator** class, specifying the type to validate and the name of the target property. This second parameter of the constructor is the validator to use to validate the property value—in this example a Regular Expression validator. Then the code can initiate validation by calling the **Validate** method, passing in the existing **ValidationResults** instance, as shown here.

```
IProduct productWithID = new Product();
PopulateInvalidProduct(productWithID);
Validator propValidator = new PropertyValueValidator<Product>("ID",
   new RegexValidator("[A-Z]{2}[0-9]{4}",
                      "Product ID must be 2 capital letters and 4 numbers.")
);
propValidator.Validate(productWithID, valResults);
```

If required, you can create a composite validator containing a combination of validators, and specify this composite validator in the second parameter. A similar technique can be used with the Field Value validator and Method Return Value validator.

After performing all of the validation operations, the example displays the results by iterating through the **ValidationResults** instance that contains the results for all of the preceding validation operations. It uses the same **ShowValidationResults** routine we described earlier in this chapter. This is the result:

```
The following 4 validation errors were detected:
+ Target object: disconnected, Member:
  - Detected by: ContainsCharactersValidator
  - Tag value: Validating the String value 'disconnected'
  - Message: 'Value must contain All of the characters 'cat'.'
+ Target object: 42, Member:
  - Detected by: DomainValidator`1[System.Int32]
  - Tag value: Validating the Integer value '42'
  - Message: 'Value must be in the list 1, 3, 7, 11, 13.'
+ Target object: MoreThan5Chars, Member:
  - Detected by: OrCompositeValidator
  - Message: 'Value can be NULL or a string of 5 characters.'
  + Nested validators:
    - Detected by: NotNullValidator
    - Message: 'Value can be NULL.'
    - Detected by: StringLengthValidator
    - Message: 'Value must be between 5 (Inclusive) and 5 (Inclusive) chars.'
+ Target object: Product, Member: ID
  - Detected by: RegexValidator
  - Message: 'Product ID must be 2 capital letters and 4 numbers.'
```

You can see how the message template tokens create the content of the messages that are displayed, and the results of the nested validators we defined for the Or Composite validator. If you want to experiment with individual validators, you can modify and extend this example routine to use other validators and combinations of validators.

WCF Service Validation Integration

This section of the chapter demonstrates how you can integrate your validation requirements for WCF services with the Validation block. The Validation block allows you to add validation attributes to the parameters of methods defined in your WCF service contract, and have the values of these automatically validated each time the method is invoked by a client.

To use WCF integration, you edit your service contract, edit the WCF configuration to add the Validation block and behaviors, and then handle errors that arise due to validation failures. In addition to the other assemblies required by Enterprise Library and the Validation block, you must add the assembly named Microsoft.Practices.EnterpriseLibrary.Validation.Integration.WCF to your application and reference them all in your service project.

The example, *Validating Parameters in a WCF Service*, demonstrates validation in a simple WCF service. It uses a service named **ProductService** (defined in the **ExampleService** project of the solution). This service contains a method named **AddNewProduct** that accepts a set of values for a product, and adds this product to its internal list of products.

Defining Validation in the Service Contract

The service contract, shown below, carries the **ValidationBehavior** attribute, and each service method defines a fault contract of type **ValidationFault**.

```
[ServiceContract]
[ValidationBehavior]
public interface IproductService
{
  [OperationContract]
  [FaultContract(typeof(ValidationFault))]
  bool AddNewProduct(
    [NotNullValidator(MessageTemplate = "Must specify a product ID.")]
    [StringLengthValidator(6, RangeBoundaryType.Inclusive,
        6, RangeBoundaryType.Inclusive,
        MessageTemplate = "Product ID must be {3} characters.")]
    [RegexValidator("[A-Z]{2}[0-9]{4}",
        MessageTemplate = "Product ID must be 2 letters and 4 numbers.")]
    string id,
    ...
    [IgnoreNulls(MessageTemplate = "Description can be NULL or a string value.")]
    [StringLengthValidator(5, RangeBoundaryType.Inclusive,
        100, RangeBoundaryType.Inclusive,
        MessageTemplate = "Description must be between {3} and {5} characters.")]
    string description,
    [EnumConversionValidator(typeof(ProductType),
        MessageTemplate = "Must be a value from the '{3}' enumeration.")]
    string prodType,
    ...
    [ValidatorComposition(CompositionType.Or,
        MessageTemplate = "Date must be between today and six months time.")]
    [NotNullValidator(Negated = true,
        MessageTemplate = "Value can be NULL or a date.")]
    [RelativeDateTimeValidator(0, DateTimeUnit.Day, 6, DateTimeUnit.Month,
        MessageTemplate = "Value can be NULL or a date.")]
    DateTime? dateDue);
}
```

You can see that the service contract defines a method named **AddNewProduct** that takes as parameters the value for each property of the **Product** class we've used throughout the examples. Although the previous listing omits some attributes to limit duplication and make it easier to see the structure of the contract, the rules applied in the example service we provide are the same as you saw in earlier examples of validating a **Product** instance. The method implementation within the WCF service is simple—it just uses the values provided to create a new **Product** and adds it to a generic **List**.

Editing the Service Configuration

After you define the service and its validation rules, you must edit the service configuration to force validation to occur. The first step is to specify the Validation block as a behavior extension. You will need to provide the appropriate version information for the assembly, which you can obtain from the configuration file generated by the configuration tool for the client application, or from the source code of the example, depending on whether you are using the assemblies provided with Enterprise Library or assemblies you have compiled yourself.

```
<extensions>
  <behaviorExtensions>
    <add name="validation"
              type="Microsoft.Practices...WCF.ValidationElement,
                    Microsoft.Practices...WCF" />
  </behaviorExtensions>
  <!-- ... other existing behavior extensions here ... -->
</extensions>
```

Next, you edit the **<behaviors>** section of the configuration to define the validation behavior you want to apply. As well as turning on validation here, you can specify a rule set name (as shown) if you want to perform validation using only a subset of the rules defined in the service. Validation will then only include rules defined in validation attributes that contain the appropriate **Ruleset** parameter (the configuration for the example application does not specify a rule set name here).

```
<behaviors>
  <endpointBehaviors>
    <behavior name="ValidationBehavior">
      <validation enabled="true" ruleset="MyRuleset" />
    </behavior>
  </endpointBehaviors>
  <!-- ... other existing behaviors here ... -->
</behaviors>
```

Note that you cannot use a configuration rule set with a WCF service—all validation rules must be in attributes.

Finally, you edit the **<services>** section of the configuration to link the **ValidationBehavior** defined above to the service your WCF application exposes. You do this by adding the **behaviorConfiguration** attribute to the service element for your service, as shown here.

```
<services>
  <service behaviorConfiguration="ExampleService.ProductServiceBehavior"
           name="ExampleService.ProductService">
    <endpoint address="" behaviorConfiguration="ValidationBehavior"
              binding="wsHttpBinding" contract="ExampleService.IProductService">
      <identity>
        <dns value="localhost" />
      </identity>
    </endpoint>
    <endpoint address="mex" binding="mexHttpBinding" contract="IMetadataExchange" />
  </service>
  <!-- ... -->
</services>
```

Using the Product Service and Detecting Validation Errors

At last you can use the WCF service you have created. The example uses a service reference added to the main project, and initializes the service using the service reference in the usual way. It then creates a new instance of a **Product** class, populates it with valid values, and calls the **AddNewProduct** method of the WCF service. Then it repeats the process, but this time by populating the product instance with invalid values. You can examine the code in the example to see this if you wish.

However, one important issue is the way that service exceptions are handled. The example code specifically catches exceptions of type **FaultException<ValidationFault>**. This is the exception generated by the service, and **ValidationFault** is the type of the fault contract we specified in the service contract.

Validation errors detected in the WCF service are returned in the **Details** property of the exception as a collection. You can simply iterate this collection to see the validation errors. However, if you want to combine them into a **ValidationResults** instance for display, especially if this is part of a multi-step process that may cause other validation errors, you must convert the collection of validation errors returned in the exception.

The example application does this using a method named **ConvertToValidationResults**, as shown here. Notice that the validation errors returned in the **ValidationFault** do not contain information about the validator that generated the error, and so we must use a null value for this when creating each **ValidationResult** instance.

```
// Convert the validation details in the exception to individual
// ValidationResult instances and add them to the collection.
ValidationResults adaptedResults = new ValidationResults();
foreach (ValidationDetail result in results)
{
  adaptedResults.AddResult(new ValidationResult(result.Message, target,
                                              result.Key, result.Tag, null));
}
return adaptedResults;
```

When you execute this example, you will see a message indicating the service being started—this may take a while the first time, and may even time out so that you need to try again. Then the output shows the result of validating the valid **Product** instance (which succeeds) and the result of validating the invalid instance (which produces the now familiar list of validation errors shown here).

```
The following 6 validation errors were detected:
+ Target object: Product, Member:
  - Detected by: [none]
  - Tag value: id
  - Message: 'Product ID must be two capital letters and four numbers.'
...
+ Target object: Product, Member:
  - Detected by: [none]
  - Tag value: description
  - Message: 'Description can be NULL or a string value.'
+ Target object: Product, Member:
  - Detected by: [none]
  - Tag value: prodType
  - Message: 'Product type must be a value from the 'ProductType' enumeration.'
...
+ Target object: Product, Member:
  - Detected by: [none]
  - Tag value: dateDue
  - Message: 'Date due must be between today and six months time.'
```

Again, we've omitted some of the duplication so that you can more easily see the result. Notice that there is no value available for the name of the member being validated or the validator that was used. This is a form of exception shielding that prevents external clients from gaining information about the internal workings of the service. However, the **Tag** value returns the name of the parameter that failed validation (the parameter names are exposed by the service), allowing you to see which of the values you sent to the service actually failed validation.

User Interface Validation Integration

You can use the Validation block to validate entire objects instead of individual controls. This makes it easier to perform validation consistently and to use more complex validation rules.

The Validation block contains integration components that make it easy to use the Validation block mechanism and rules to validate user input within the user interface of ASP.NET, Windows Forms, and WPF applications. While these technologies do include facilities to perform validation, this validation is generally based on individual controls and values.

When you integrate the Validation block with your applications, you can validate entire objects, and collections of objects, using sets of rules you define. You can also apply complex validation using the wide range of validators included with the Validation block. This allows you to centrally define a single set of validation rules, and apply them in more than one layer and when using different UI technologies.

> *The UI integration technologies provided with the Validation block do not instantiate the classes that contain the validation rules. This means that you cannot use self-validation with these technologies.*

ASP.NET User Interface Validation

The Validation block includes the **PropertyProxyValidator** class that derives from the ASP.NET **BaseValidator** control, and can therefore take part in the standard ASP.NET validation cycle. It acts as a wrapper that links an ASP.NET control on your Web page to a rule set defined in your application through configuration, attributes, and self-validation.

To use the **PropertyProxyValidator**, you must install the NuGet package named **EnterpriseLibrary.Validation.Integration.AspNet**. You must also include a **Register** directive in your Web pages to specify this assembly and the prefix for the element that will insert the **PropertyProxyValidator** into your page.

```
<% @Register TagPrefix="EntLibValidators"
Assembly="Microsoft.Practices.EnterpriseLibrary.Validation.Integration.AspNet"
Namespace="Microsoft.Practices.EnterpriseLibrary.Validation.Integration.AspNet"
%>
```

Then you can define the validation controls in your page. The following shows an example that validates a text box that accepts a value for the **FirstName** property of a **Customer** class, and validates it using the rule set named **RuleSetA**.

```
<EntLibValidators:PropertyProxyValidator id="firstNameValidator"
   runat="server" ControlToValidate="firstNameTextBox"
   PropertyName="FirstName" RulesetName="RuleSetA"
   SourceTypeName="ValidationQuickStart.BusinessEntities.Customer" />
```

One point to be aware of is that, unlike the ASP.NET validation controls, the Validation block **PropertyProxyValidator** control does not perform client-side validation. However, it does integrate with the server-based code and will display validation error messages in the page in the same way as the ASP.NET validation controls.

For more information about ASP.NET integration, see the *Enterprise Library Reference Documentation*.

Windows Forms User Interface Validation

The Validation block includes the **ValidationProvider** component that extends Windows Forms controls to provide validation using a rule set defined in your application through configuration, attributes, and self-validation. You can handle the **Validating** event to perform validation, or invoke validation by calling the **PerformValidation** method of the control. You can also specify an **ErrorProvider** that will receive formatted validation error messages.

To use the **ValidationProvider**, you add the assembly named Microsoft.Practices.EnterpriseLibrary.Validation.Integration.WinForms to your application, and reference it in your project.

For more information about Windows Forms integration, see the *Enterprise Library Reference Documentation*.

WPF User Interface Validation

The Validation block includes the **ValidatorRule** component that you can use in the binding of a WPF control to provide validation using a rule set defined in your application through configuration, attributes, and self-validation. To use the **ValidatorRule**, you add the assembly named Microsoft.Practices.EnterpriseLibrary.Validation.Integration.WPF to your application, and reference it in your project.

As an example, you can add a validation rule directly to a control, as shown here.

```
<TextBox x:Name="TextBox1">
  <TextBox.Text>
    <Binding Path="ValidatedStringProperty" UpdateSourceTrigger="PropertyChanged">
      <Binding.ValidationRules>
        <vab:ValidatorRule SourceType="{x:Type test:ValidatedObject}"
                           SourcePropertyName="ValidatedStringProperty"/>
      </Binding.ValidationRules>
    </Binding>
  </TextBox.Text>
</TextBox>
```

You can also specify a rule set using the **RulesetName** property, and use the **ValidationSpecificationSource** property to refine the way that the block creates the validator for the property.

For more information about WPF integration, see the *Enterprise Library Reference Documentation*.

Creating Custom Validators

While the wide range of validators included with the Validation block should satisfy most requirements, you can easily create your own custom validators and integrate them with the block. This may be useful if you have some specific and repetitive validation task that you need to carry out, and which is more easily accomplished using custom code.

The easiest way to create a custom validator is to create a class that inherits from one of the abstract base classes provided with the Validation block. Depending on the type of validation you need to perform, you may choose to inherit from base types such as the **ValueValidator** or **MemberAccessValidator** classes, the **Validator<T>** base class (for a strongly typed validator) or from the **Validator** class (for a loosely typed validator).

You can also create your own custom validation attributes that will apply custom validators you create. The base class, **ValidatorAttribute**, provides a good starting point for this.

For more information on extending Enterprise Library and creating custom providers, see the *Enterprise Library Reference Documentation*.

Summary

In this chapter we have explored the Enterprise Library Validation block and shown you how easy it is to decouple your validation code from your main application code. The Validation block allows you to define validation rules and rule sets; and apply them to objects, method parameters, properties, and fields of objects you use in your application. You can define these rules using configuration, attributes, or even using custom code and self-validation within your classes.

Validation is a vital crosscutting concern, and should occur at the perimeter of your application, at trust boundaries, and (in most cases) between layers and distributed components. Robust validation can help to protect your applications and services from malicious users and dangerous input (including SQL injection attacks); ensure that it processes only valid data and enforces business rules; and improve responsiveness.

The ability to centralize your validation mechanism and the ability to define rules through configuration also make it easy to deploy and manage applications. Administrators can update the rules when required without requiring recompilation, additional testing, and redeployment of the application. Alternatively, you can define rules, validation mechanisms, and parameters within your code if this is a more appropriate solution for your own requirements.

More Information

All links in this book are accessible from the book's online bibliography on MSDN at *http://aka.ms/el6biblio*.

For more details on each validator, see the *Enterprise Library Reference Documentation*.

For more information on using interception, see "*Chapter 5 - Interception using Unity*" in the Developer's Guide to Dependency Injection Using Unity.

For more information, see the *Enterprise Library Reference Documentation*, and the topic "*System.ComponentModel.DataAnnotations Namespace*" on MSDN.

The topic, "*Using Self Validation*" in the Enterprise Library Reference Documentation, demonstrates use of the Validation block attributes.

For more information about data annotations, see "*System.ComponentModel.DataAnnotations Namespace*" on MSDN.

For more information about ASP.NET integration, see the *Enterprise Library Reference Documentation*.

For more information about Windows Forms integration, see the *Enterprise Library Reference Documentation*.

For more information about WPF integration, see the *Enterprise Library Reference Documentation*.

For more information on extending Enterprise Library and creating custom providers, see the *Enterprise Library Reference Documentation*.

General Links:

- There are resources to help if you're getting started with Enterprise Library, and there's help for existing users as well (such as the breaking changes and migration information for previous versions) available on the Enterprise Library Community Site at *http://www.codeplex.com/entlib/*.
- For more details about the Enterprise Library Application Blocks, see the *Enterprise Library Reference Documentation* and the *Enterprise Library 6 Class Library*.
- You can download and work through the Hands-On Labs for Enterprise Library, which are available at *http://aka.ms/el6hols*.
- You can download the Enterprise Library binaries, source code, Reference Implementation, or QuickStarts from the Microsoft Download Center at *http://go.microsoft.com/fwlink/p/?LinkID=290898*.
- To help you understand how you can use Enterprise Library application blocks, we provide a series of simple example applications that you can run and examine. To obtain the example applications, go to *http://go.microsoft.com/fwlink/p/?LinkID=304210*.

8 Updating aExpense to Enterprise Library 6

This chapter describes how the aExpense reference implementation was upgraded from Enterprise Library 5 to Enterprise Library 6. The upgrade process included replacing functionality that was removed from Enterprise Library 6, updating the application to work with the new versions of existing application blocks, and modifying the application to take advantage of some of the new features in Enterprise Library 6. The updated version of aExpense also takes advantage of some of the features in the .NET Framework 4.5, such as caching and the use of claims.

The chapter describes the before and after state of the implementation as it relates to each of the Enterprise Library blocks that aExpense uses. It also highlights any issues encountered, significant changes made, and decisions taken by the team in relation to the update.

The aExpense Application

The aExpense application allows employees at Adatum (a fictional company) to submit, track, and process business expenses. Everyone in Adatum uses this application to request reimbursements.

The application is representative of many other applications in Adatum's portfolio so it's a good test case for upgrading to Enterprise Library 6. The developers at Adatum hope to gain a detailed understanding of what it takes to upgrade an application that uses Enterprise Library 5.0 to Enterprise Library 6 from this project before embarking on similar upgrades to some of the larger, more critical, systems at Adatum.

The aExpense application is typical of many LOB applications that use Enterprise Library.

Adatum also upgraded the aExpense application from using version 4 of the .NET Framework to version 4.5 of the .NET Framework as part of the migration project.

The aExpense Architecture

Figure 1 illustrates the architecture of the version of aExpense that uses Enterprise Library 5. The sample application simulates the use of Active Directory. In the version of aExpense that uses Enterprise Library 6, events are logged to file and database tables, not to the Windows Event Log.

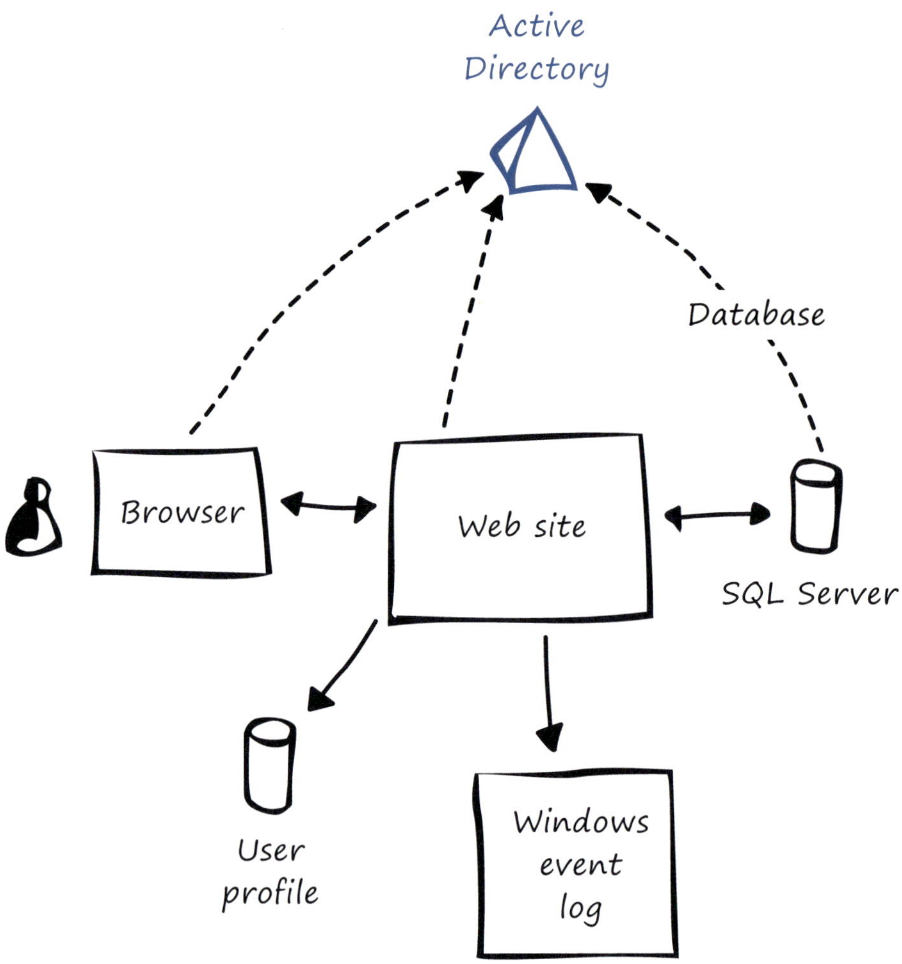

FIGURE 1
aExpense architecture

Like many of Adatum's applications, the aExpense application uses Enterprise Library and Unity.

The architecture is straightforward and one that many other applications use. aExpense is an ASP.NET application and employees use a browser to interact with it. The application makes use of many of the Enterprise Library 5.0 blocks and uses Unity as a dependency injection container.

The application simulates using Windows authentication for security. To store user preferences, it relies on ASP.NET membership and profile providers. Exceptions and logs are implemented with Enterprise Library's Exception Handling Application Block and Logging Application Block. The website simulates using Directory Services APIs to query for employee data stored in Active Directory, such as the employee's manager. Managers can approve the expenses.

The aExpense application implements the trusted subsystem to connect to SQL Server. It authenticates with a Windows domain account. The SQL database uses integrated authentication mode. The aExpense application stores its information on SQL Server.

Adatum's aExpense application uses a standard website architecture based on ASP.NET with data stored in SQL Server. However, it does integrate with other in-house systems.

Using Enterprise Library v5.0

The aExpense application uses many of the application blocks from Enterprise Library 5.0. This section describes, briefly, how and why aExpense uses these blocks before migrating to version 6. You may find it useful to have the version of aExpense that uses Enterprise Library v5.0 open in Visual Studio as you read through this section. This will enable you to explore the application in greater detail.

Caching

The aExpense application makes use of the Caching Application Block in the **ExpenseRepository** class. It caches expense items when it retrieves them from the backing store to minimize round-trip times on subsequent requests for the same data: you can see this behavior in the **GetExpensesForApproval**, **GetExpensesById**, **GetExpesnsesByUSer**, and **UpdateCache** methods. The **Initialize** method in this class reads the configuration file and initializes the cache. The block also encrypts the cache contents.

Encrypting the cache contents accounts for the use of the "Enterprise Library 5.0 – Caching Application Block Cryptography Provider" and "Enterprise Library 5.0 – Cryptography Application Block" NuGet packages.

Logging

The aExpense application makes extensive use of the version 5.0 Logging Application Block to write log data to a variety of locations. If you examine the logging configuration in the Web.EnterpriseLibrary.config file, you will see four different listeners defined. One of these listeners is a database trace listener, and this accounts for the use of the version 5.0 "Logging Application Block Database Provider" and "Data Access Application Block" NuGet packages.

> *The **SimulatedLdapProfileStore** class also uses a **Database** instance from the Data Access Application Block.*

In the aExpense application, the **TracingBehavior** class shown below is responsible for writing to the log. This class defines a virtual method interceptor that is configured in the Unity container; you can see how the interceptor is configured in the **ContainerBootstrapper** class. Notice that this interceptor class is used with the **Model.User** class.

```
public class TracingBehavior : IInterceptionBehavior
{
  public IMethodReturn Invoke(IMethodInvocation input, GetNextInterceptionBehaviorDelegate getNext)
  {
    if (Logger.IsLoggingEnabled())
    {
      // Only log methods with arguments (bypass getters)
      if (input.Arguments.Count > 0)
      {
        string arguments = string.Join(",", input.Arguments.OfType<object>());
        Logger.Write(string.Format(
                                  CultureInfo.InvariantCulture,
                                  "{0}: Method {1}.{2} executed. Arguments: {3}",
                                  DateTime.Now,
                                  input.MethodBase.DeclaringType.Name,
                                  input.MethodBase.Name,
                                  arguments), Constants.ExpenseTracingCategory,
                                  Constants.TracingBehaviorPriority);
      }
    }

    return getNext()(input, getNext);
  }

  public IEnumerable<Type> GetRequiredInterfaces()
  {
    return Enumerable.Empty<Type>();
  }

  public bool WillExecute
  {
    get { return true; }
  }
}
```

You can see the messages written by this behavior in the Logging database in the Log table. For example: "04/18/2013 11:27:41: Method User.set_UserName executed"

Exception Handling

The configuration of the Exception Handling Application Block defines two exception policies: "Notify Policy" and "Global Policy." Both of these policies make use of the Exception Handling Application Block Logging Handler to write details of any exceptions to the logging database.

You can see examples of the Exception Handling block in use in the **ExpenseRepository** and **Global** classes.

Validation

The aExpense application uses the Validation Application Block attributes in the model classes **Expense** and **ExpenseItem** to enforce validating the properties defined in these classes. You can also see the "Validation Application Block Integration with ASP.NET" NuGet package in use in the **AddExpense.aspx** file: Look for the **<cc1:PropertyProxyValidator>** tags.

The application uses two custom validators. The **ApproverValidator** class validates the approver name in the **Expense** class, and the **DuplicateExpenseDetailsValidator** class is part of the **ExpenseRulset** defined in the configuration file.

Security

The aExpense application uses the Authorization Rule Provider from the version 5.0 Security Application Block to determine whether the current user is authorized to approve an expense. The **UpdateApproved** method in the **ExpenseRepository** class is decorated with the custom **RulesPrincipalPermission** attribute. This attribute class uses the **RulesPrincipalPermission** class, which in turn reads the authorization rule from the configuration file.

Policy Injection

The aExpense application uses policy injection to log saving expenses in the **ExpenseRepository** class. The **SaveExpense** method has an attribute **Tag("SaveExpensePolicyRule")**. The configuration file contains the policy definition, which is shown below.

```
<policyInjection>
  <policies>
    <add name="ExpenseRepositoryTracingPolicy">
      <matchingRules>
        <add name="TagRule" match="SaveExpensePolicyRule"
             ignoreCase="false"
             type="Microsoft.Practices.EnterpriseLibrary.PolicyInjection
                  .MatchingRules.TagAttributeMatchingRule, ..."/>
      </matchingRules>
      <handlers>
        <add type="Microsoft.Practices.EnterpriseLibrary.Logging
                   .PolicyInjection.LogCallHandler, ..."
             logBehavior="After"
             beforeMessage="Before invoking"
             afterMessage="After invoking"
             name="Logging Call Handler">
          <categories>
            <add name="ExpenseTracing" />
          </categories>
        </add>
      </handlers>
    </add>
  </policies>
</policyInjection>
```

You can see log entries created by the Logging Call Handler in the Logging database in the Log table by searching for log entries with a title "Call Logging."

Unity

The **ContainerBootstrapper** class includes the following code.

```
container
    .AddNewExtension<EnterpriseLibraryCoreExtension>()
    .RegisterType<IProfileStore, SimulatedLdapProfileStore>(
      new ContainerControlledLifetimeManager())
    .RegisterType<IUserRepository, UserRepository>(
      new ContainerControlledLifetimeManager())
    .RegisterType<AExpense.Model.User>(
      new Interceptor<VirtualMethodInterceptor>(),
      new InterceptionBehavior<TracingBehavior>())
    .RegisterType<IExpenseRepository, ExpenseRepository>(
      new ContainerControlledLifetimeManager(),
      new Interceptor<VirtualMethodInterceptor>(),
      new InterceptionBehavior<PolicyInjectionBehavior>());

UnityServiceLocator locator = new UnityServiceLocator(container);
ServiceLocator.SetLocatorProvider(() => locator);
```

This code adds the Enterprise Library extension to the container that enables you to resolve types defined in the Enterprise Library blocks in your code. It also registers a number of application types and configures the Unity service locator.

The **Approve** page class uses the **Dependency** attribute to indicate that an **IExpenseRepository** instance should be injected into the instance. This **BuildUp** method performs this injection in the **Application_PreRequestHandlerExecute** method in the **Global** class as shown in the following code sample.

```
protected void Application_PreRequestHandlerExecute(object sender, EventArgs e)
{
    ...
    Page handler = HttpContext.Current.Handler as Page;
    if (handler != null)
    {
        IUnityContainer container = Application.GetContainer();
        container.BuildUp(handler.GetType(), handler);
    }
}
```

NuGet Packages and References – Before and After

As an overview of how aExpense uses Enterprise Library 5.0 before the update, the following table lists the Enterprise Library NuGet packages and related references in the aExpense application.

NuGet Package	References
Unity (version 2.1)	Microsoft.Practices.Unity Microsoft.Practices.Unity.Configuration
Unity Interception Extension (version 2.1)	Microsoft.Practices.Unity.Interception Microsoft.Practices.Unity.Interception.Configuration
CommonServiceLocator	Microsoft.Practices.ServiceLocation

NuGet Package	References
Enterprise Library 5.0 – Common Infrastructure	Microsoft.Practices.EnterpriseLibrary.Common
Enterprise Library 5.0 – Caching Application Block	Microsoft.Practices.EnterpriseLibrary.Caching
Enterprise Library 5.0 – Caching Application Block Cryptography Provider	Microsoft.Practices.EnterpriseLibrary.Caching.Cryptography
Enterprise Library 5.0 – Cryptography Application Block	Microsoft.Practices.EnterpriseLibrary.Security.Cryptography
Enterprise Library 5.0 – Data Access Application Block	Microsoft.Practices.EnterpriseLibrary.Data
Enterprise Library 5.0 – Exception Handling Application Block	Microsoft.Practices.EnterpriseLibrary.ExceptionHandling
Enterprise Library 5.0 – Exception Handling Application Block Logging Handler	Microsoft.Practices.EnterpriseLibrary.ExceptionHandling.Logging
Enterprise Library 5.0 – Logging Application Block	Microsoft.Practices.EnterpriseLibrary.Logging
Enterprise Library 5.0 – Logging Application Block Database Provider	Microsoft.Practices.EnterpriseLibrary.Logging.Database
Enterprise Library 5.0 – Policy Injection Application Block	Microsoft.Practices.EnterpriseLibrary.PolicyInjection
Enterprise Library 5.0 – Security Application Block	Microsoft.Practices.EnterpriseLibrary.Security
Enterprise Library 5.0 – Validation Application Block	Microsoft.Practices.EnterpriseLibrary.Validation
Enterprise Library 5.0 – Validation Application Block Integration with ASP.NET	Microsoft.Practices.EnterpriseLibrary.Validation.Integration.AspNet

The following table shows the NuGet packages and references used in the aExpense application after the update.

NuGet Package	References
Unity (version 3)	Microsoft.Practices.Unity Microsoft.Practices.Unity.Configuration
Unity Interception Extension (version 3)	Microsoft.Practices.Unity.Interception Microsoft.Practices.Unity.Interception.Configuration
CommonServiceLocator	Microsoft.Practices.ServiceLocation
Enterprise Library 6 – Data Access Application Block	Microsoft.Practices.EnterpriseLibrary.Data
Enterprise Library 6 – Exception Handling Application Block	Microsoft.Practices.EnterpriseLibrary.ExceptionHandling
Enterprise Library 6 – Policy Injection Application Block	Microsoft.Practices.EnterpriseLibrary.PolicyInjection
Enterprise Library 6 – Semantic Logging Application Block (version 1.0)	Microsoft.Practices.EnterpriseLibrary.SemanticLogging
Enterprise Library 6 – Semantic Logging Application Block – SQL Server Sink (version 1.0)	Microsoft.Practices.EnterpriseLibrary.SemanticLogging.Database
Enterprise Library 6 – Transient Fault Handling Application Block	Microsoft.Practices.EnterpriseLibrary.TransientFaultHandling Microsoft.Practices.EnterpriseLibrary.TransientFaultHandling.Configuration
Enterprise Library 6 – Transient Fault Handling Application Block – Windows Azure SQL Database integration	Microsoft.Practices.EnterpriseLibrary.TransientFaultHandling.Data
Enterprise Library 6 – Validation Application Block	Microsoft.Practices.EnterpriseLibrary.Validation
Enterprise Library 6 – Validation Application Block Integration with ASP.NET	Microsoft.Practices.EnterpriseLibrary.Validation.Integration.AspNet

Handling Blocks Removed from Enterprise Library 6

You can use the PowerShell script, **Update-EntlibConfiguration.ps1**, to update the version numbers in all of the configuration files in your project. It also updates any changed namespaces in the configuration files. You can find this script in the [Solution Folder]\packages\EnterpriseLibrary.Common.[Version] folder after you install an Enterprise Library 6 NuGet package.

The version of aExpense that uses the Enterprise Library 5.0 blocks uses both the Caching Application Block and the Security Application Block, neither of which are part of Enterprise Library 6. The developers at Adatum had to replace the functionality used from these blocks in the latest version of the aExpense application.

Replacing Caching Functionality

In the new version of aExpense that uses Enterprise Library 6, classes in the **System.Runtime.Caching** namespace provide the caching functionality for the **ExpenseRepository** class. The changes to the **GetExpensesForApproval**, **GetExpensesById**, **GetExpesnsesByUSer**, and **UpdateCache** methods which access the cache are minimal. The **Initialize** method now configures the cache using a cache instance from the System.Runtime.Caching namespace.

To make it easier to install, the aExpense RI currently uses an in-memory caching solution that does not support distributed caching scenarios such as web farm deployments. For a distributed caching solution, you should consider using AppFabric for Windows Server, which requires minimal code changes in the application. For more information, see "*AppFabric Caching API Usage Sample*" on MSDN.

Replacing the Authorization Rule Provider

The new version of aExpense now uses claims-based authorization to determine role membership. The **UpdateApproved** method in the **ExpenseRepository** class is now decorated with the **ClaimsPrincipalPermission** attribute. aExpense also authorizes page access through settings on the **claimsAuthorizationManager** section of the configuration file.

The RI shows a very basic implementation of claims based authorization. You can read more about claims-based authentication and authorization in the guide "*A Guide to Claims-Based Identity and Access Control (2nd Edition)*" available for download from MSDN.

Handling Blocks Updated in Enterprise Library 6

The version of aExpense that uses Enterprise Library 6, continues to use the Exception Handling Application Block, the Validation Application Block, the Policy Injection Application Block, and Unity. There are some changes in the way that aExpense configures and uses these blocks that are discussed in this section.

Using the Exception Handling Application Block Version 6

The major change in the use of the Exception Handling block is due to a change in the logging approach used by aExpense; it now uses the Semantic Logging Application Block in place of the Logging Application Block. This change is discussed later in this chapter. You can see the impact of this in the Web.EnterpriseLibrary.config file in the **exceptionHandlingSection** where the old "Logging Exception Handler" has been replaced with a custom "SemanticLogging Exception Handler" definition.

There are also changes in the **ContainerBootstrapper** class that relate to the Exception Handling Application Block because the **ExceptionManager** type is not registered to the Unity container automatically. These changes are described in detail in the section covering Unity later in this chapter.

The version numbers in the configuration file have been updated to reflect the new Enterprise Library version.

Both versions of the aExpense application resolve the **ExceptionManager** type from the container in the **ExpenseRepository** and **UserRepository** classes in the same way. However, because version 6 of the block does not automatically configure the static **ExceptionPolicy** class, the developers at Adatum chose to modify the **Global** class to resolve the **ExceptionManager** type from the container used throughout the application as shown in the following code sample.

```
protected void Application_Start(object sender, EventArgs e)
{
    ...

    IUnityContainer container = new UnityContainer();
    ContainerBootstrapper.Configure(container);

    Application.SetContainer(container);

    ...
}

private void Application_Error(object sender, EventArgs e)
{
    Exception ex = Server.GetLastError();
    ...
    var exceptionManager = Application.GetContainer().Resolve<ExceptionManager>();
    if (exceptionManager != null)
    {
        exceptionManager.HandleException(ex, Constants.GlobalPolicy);
    }
}
```

Using the Validation Application Block Version 6

There are no changes in the way that the aExpense application uses the Validation Application Block. All of the existing validation is done either using attributes or using the "Validation Application Block Integration with ASP.NET" NuGet package and tags in the .aspx files.

Using the Policy Injection Application Block Version 6

The **ContainerBootstrapper** class now includes code to load the Policy Injection Block settings from the configuration file and use them to configure the container. Previously, in Enterprise Library 5.0, this happened automatically. The following code sample shows how to do this in version 6.

```
var policyInjectionSettings = (PolicyInjectionSettings)source.
                              GetSection(PolicyInjectionSettings.SectionName);
policyInjectionSettings.ConfigureContainer(container);
```

The aExpense solution that uses Enterprise Library 5.0 used the logging call handler that was included with Enterprise Library. However, the aExpense solution that uses Enterprise Library 6 uses the Semantic Logging Application Block in place of the Logging Application Block. It therefore includes a custom call handler for semantic logging. You can see the updated configuration for the block in the Web.EnterpriseLibrary.config file. The following code sample shows the custom **SemanticLogCallHandler** class included in the solution that handles logging for the expense tracing policy.

```
[ConfigurationElementType(typeof(CustomCallHandlerData))]
public class SemanticLogCallHandler : IcallHandler
{
    public SemanticLogCallHandler(NameValueCollection attributes)
    {
    }

    public IMethodReturn Invoke(IMethodInvocation input,
                                GetNextHandlerDelegate getNext)
    {
        if (getNext == null) throw new ArgumentNullException("getNext");

        AExpenseEvents.Log.LogCallHandlerPreInvoke(
          input.MethodBase.DeclaringType.FullName,
          input.MethodBase.Name);

        var sw = Stopwatch.StartNew();

        IMethodReturn result = getNext()(input, getNext);

        AExpenseEvents.Log.LogCallHandlerPostInvoke(
          input.MethodBase.DeclaringType.FullName,
          input.MethodBase.Name, sw.ElapsedMilliseconds);

        return result;
    }

    public int Order { get; set; }
}
```

Using Unity Version 3

To simplify the registration types in the Unity container, the developers at Adatum replaced the individual registrations of the **ExpenseRepository**, **SimulatedLdapProfileStore**, and **UserRepository** types with the following code in the **ContainerBootStrapper** class.

```
container.RegisterTypes(
  AllClasses.FromAssemblies(Assembly.GetExecutingAssembly()),
  WithMappings.FromAllInterfacesInSameAssembly,
  WithName.Default,
  WithLifetime.ContainerControlled);
```

This approach, using the new registration by convention feature in Unity, provides minimal benefits in this scenario because the application currently only registers three types. However, as the application grows and becomes more complex this approach becomes more useful.

The Enterprise Library blocks no longer have a dependency on Unity, and in consequence, they no longer bootstrap a Unity container automatically. The aExpense solution that uses Enterprise Library 5.0 includes the following code to register the blocks and configure them based on the configuration data in the configuration file.

```
container
  .AddNewExtension<EnterpriseLibraryCoreExtension>();
```

To minimize the changes in the aExpense solution that uses Enterprise Library 6, the developers at Adatum chose to register the blocks the application uses manually in the container as shown in the following code sample. In this way, the developers don't have to make any changes elsewhere in the application where the Enterprise Library types are resolved.

```
IConfigurationSource source = ConfigurationSourceFactory.Create();

var policyInjectionSettings = (PolicyInjectionSettings)source
     .GetSection(PolicyInjectionSettings.SectionName);
policyInjectionSettings.ConfigureContainer(container);

var policyFactory = new ExceptionPolicyFactory(source);
var dbFactory = new DatabaseProviderFactory(source);
var validationFactory =
     ConfigurationValidatorFactory.FromConfigurationSource(source);

container
     .RegisterType<ExceptionManager>(
        new InjectionFactory(c => policyFactory.CreateManager()))
     .RegisterType<Database>(
        new InjectionFactory(c => dbFactory.CreateDefault()))
     .RegisterInstance<ValidatorFactory>(validationFactory);
```

You don't need to register the Enterprise Library types with the container, you can always use the factories to instantiate the objects directly. However, because the aExpense application was originally designed with Enterprise Library 5.0 in mind, it resolves the Enterprise Library types from the container in many places throughout the application, so registering the necessary types resulted in fewer changes, and all the changes are centralized in the **ContainerBootstrapper** class.

Using the New Blocks in Enterprise Library Version 6

The new version of the aExpense application uses both new Enterprise Library 6 blocks: the Semantic Logging Application Block, and the Transient Fault Handling Application Block. The developers at Adatum decided to replace the existing logging in the application that was based on the Logging Application Block with the Semantic Logging Application Block in order to gain some of the benefits associated with this logging approach. They added the Transient Fault Handling Application Block: this is a first step to prepare the application for using SQL Database running in the cloud in place of the existing on-premises database solution.

The Semantic Logging Application Block

The Semantic Logging Application Block enables Adatum to collect semantic/structured logging information from the aExpense application. Structured log information will enable support staff to query and analyze the log data collected from the aExpense application. If Adatum chooses to host the Semantic Logging Application Block in a separate Windows service, Adatum can make the logging process more robust: a major failure in the aExpense application is less likely to lose logging data at the time the crash occurs, making troubleshooting more effective in analyzing the cause of the problem.

The new version of the aExpense application includes the **AExpenseEvents** class that derives from the **EventSource** class in the **System.Diagnostics.Tracing** namespace. This class defines all of the messages that the aExpense application can generate. It also defines various pieces of metadata associated with trace messages such as keywords, opcodes, and tasks. The application can generate a trace message by invoking any of the message methods in this class. For example, the **SaveExpense** method in the **ExpenseRepository** class records when a save operation begins and ends as shown in the following code sample.

```
public virtual void SaveExpense(Model.Expense expense)
{
    exManager.Process(() =>
    {
        AExpenseEvents.Log.SaveExpenseStarted(expense.Id, expense.Title);
        ProcessInContext((db) =>
        {
            ...
        });

        ...
        AExpenseEvents.Log.SaveExpenseFinished(expense.Id, expense.Title);
    },
    ...);
}
```

Notice how the calling code doesn't need to provide information such as the log level or category. That's all defined in the custom **EventSource** class.

By default, the aExpense application uses the Semantic Logging Application Block in the in-process mode. If you want to use the Semantic Logging Application Block in its out-of-process mode, you must install and configure the "Enterprise Library Semantic Logging Service" host to collect, filter, and save the log messages from aExpense. The Visual Studio solution includes a folder containing a configuration file (SemanticLogging-svc.xml) for the service and a readme file describing how to set it up.

The service configuration file defines the following sinks for log messages:

- **DataAccess**. A flat file destination that receives all log messages tagged with the **DataAccess** keyword in the **AExpense** class.
- **UserInterface**. A rolling flat file destination that receives all log messages tagged with the **UserInterface** keyword and a severity of **Informational** or higher in the **AExpense** class.
- **SQL-All**. A SQL Server database destination that receives all log messages.

The Transient Fault Handling Application Block

Although the aExpense application currently uses an on premises SQL Server solution, Adatum plans to migrate the application's data to SQL Database running in the cloud. In preparation for this migration, the developers at Adatum added code to the solution to manage retries for the transient errors than can occur when accessing SQL Database.

The SQL detection strategy in the Transient Fault Handling Application Block targets SQL Database running in the cloud rather than SQL Server running on premises. When Adatum changes the connection strings in the application to connect the application to the cloud-based SQL Database, the retry policies will add greater resilience to the calls to the database.

The retry configuration is stored in the Web.EnterpriseLibrary.config file along with the other Enterprise Library block configuration data and defines a single exponential back off policy. As with the other blocks, the **ContainerBootstrapper** class contains code to load this configuration data and initialize the factory as shown in the following code sample.

```
var retryPolicySettings =
    RetryPolicyConfigurationSettings.GetRetryPolicySettings(source);
RetryPolicyFactory.SetRetryManager(
    retryPolicySettings.BuildRetryManager(), throwIfSet: false);
```

The aExpense application uses the retry policy in the **ExpenseRepository** and **SimulatedLdapProfileStore** classes to wrap the calls to the database. The following is an example from the **ExpenseRepository** class.

```
private void Initialize(...)
{
  ...
  this.retryPolicy = RetryPolicyFactory
            .GetRetryPolicy<SqlDatabaseTransientErrorDetectionStrategy>();
}
Public virtual void SaveExpense(Model.Expense expense)
{
 ...
   this.retryPolicy.ExecuteAction(() => db.SubmitChanges());
 ...
}
```

Although the code to initialize the **RetryPolicyFactory** object is in the **ContainerBootstrapper** class, it is not using Unity. There is no preexisting code that resolves retry policies from the container; therefore, the developers chose to use the static facades in the block.

Other Possible Changes

The changes illustrated in these two versions of the aExpense application represent the basic changes the developers at Adatum needed to make to migrate the application to Enterprise Library 6, the replacement of the Logging Application Block with the Semantic Logging Application Block, and the introduction of the Transient Fault Handling Application Block. There are additional changes that could be introduced to realize some additional benefits from Enterprise Library 6 such as:

- Use programmatic configuration in place of declarative configuration. Many of the settings in the configuration file do not need to be changed after the application is deployed: by moving them to programmatic configuration, they would no longer be exposed.

- Increased use of static facades. Adatum chose to register types from the Database Access Application Block, the Exception Handling Application Block, and Validation Application Block in the Unity container in order to minimize the impact of the migration in other areas of the application. As an alternative, Adatum could follow the pattern adopted by the Transient Fault Handing Application Block and use static facades instead.

More Information

All links in this book are accessible from the book's online bibliography on MSDN at *http://aka.ms/el6biblio*.

For more information about Unity, see the *Dependency Injection with Unity* guide. This guide is also available on MSDN: *Developer's Guide to Dependency Injection Using Unity*.

For more information about migrating to Enterprise Library 6, see the *Migration Guide*.

For more information on distributed caching solutions, see "*AppFabric Caching API Usage Sample*" on MSDN.

You can read more about claims-based authentication and authorization in the guide "*A Guide to Claims-Based Identity and Access Control (2nd Edition)*" available for download from MSDN.

General Links:

- There are resources to help if you're getting started with Enterprise Library, and there's help for existing users as well (such as the breaking changes and migration information for previous versions) available on the Enterprise Library Community Site at *http://www.codeplex.com/entlib/*.
- For more details about the Enterprise Library Application Blocks, see the *Enterprise Library Reference Documentation* and the *Enterprise Library 6 Class Library*.
- You can download and work through the Hands-On Labs for Enterprise Library, which are available at *http://aka.ms/el6hols*.
- You can download the Enterprise Library binaries, source code, Reference Implementation, or QuickStarts from the Microsoft Download Center at *http://go.microsoft.com/fwlink/p/?LinkID=290898*.
- To help you understand how you can use Enterprise Library application blocks, we provide a series of simple example applications that you can run and examine. To obtain the example applications, go to *http://go.microsoft.com/fwlink/p/?LinkID=304210*.

Tales from the Trenches

First Experiences

My first experiences with the Logging and Exception Handling Blocks

Contributed by Fabian Fernandez

Logging and exception handling are two things you always have in mind when starting to build a new project. I was really tired of writing new implementations for these every time I started a project because of the different requirements tied to the specifics of each project. Therefore, when I was the .NET Software Architect at a previous job starting a project from zero, I decided to search for a one-time solution to cover these areas. A solution that would, when we needed to reuse it in another project or satellite application, work like magic with just a couple of configuration changes and as few lines of code as possible.

I had heard of Enterprise Library but never used it, so I began by looking at the documentation, starting with the EntLib50.chm help file that you can find on *CodePlex*. My first impression was that it is initially a little hard to understand because of the quantity of information, but after you start reading sections, it becomes like a walk in the park. It takes just a couple of hours to understand the basics.

The documentation is also available on MSDN for *Enterprise Library 5* and *Enterprise Library 6*.

The first thing I decided to use was the ability to redirect sections in the configuration file using the **FileConfigurationSource** type; this gives me the ability to reuse both my logging and exception handling assemblies and their associated configuration with my application specific code. I end up using two assemblies, one for logging, and one for exception handling, each with its own configuration file. When I reference one of the assemblies, I include the configuration file as a linked file so I don't need to make multiple copies of the configuration file all over the place: linked files are powerful, use them!

Here's an example of a web.config/app.config using the redirect sections feature to reuse the configuration file of each block:

```
<?xml version="1.0" encoding="utf-8" ?>
<configuration>
  <configSections>
    <section name="enterpriseLibrary.ConfigurationSource"
             type="Microsoft.Practices.EnterpriseLibrary.Common.Configuration
                  .ConfigurationSourceSection, Microsoft.Practices.EnterpriseLibrary
                  .Common, Version=5.0.505.0, Culture=neutral,
                  PublicKeyToken=31bf3856ad364e35" requirePermission="true" />
  </configSections>
```

```
    <enterpriseLibrary.ConfigurationSource
      selectedSource="System Configuration Source">
      <sources>
        <add name="System Configuration Source" type="Microsoft.Practices
          .EnterpriseLibrary.Common.Configuration.SystemConfigurationSource,
          Microsoft.Practices.EnterpriseLibrary.Common, Version=5.0.505.0,
          Culture=neutral, PublicKeyToken=31bf3856ad364e35" />
        <add name="Logging configuration" type="Microsoft.Practices
          .EnterpriseLibrary.Common.Configuration.FileConfigurationSource,
          Microsoft.Practices.EnterpriseLibrary.Common, Version=5.0.505.0,
          Culture=neutral, PublicKeyToken=31bf3856ad364e35"
          filePath="./Configs/EnterpriseLibrary.Logging.config" />
        <add name="Exception Handling configuration" type="Microsoft.Practices
          .EnterpriseLibrary.Common.Configuration.FileConfigurationSource,
          Microsoft.Practices.EnterpriseLibrary.Common, Version=5.0.505.0,
          Culture=neutral, PublicKeyToken=31bf3856ad364e35"
          filePath="./Configs/EnterpriseLibrary.ExceptionHandling.config" />
      </sources>
      <redirectSections>
        <add sourceName="Logging configuration" name="loggingConfiguration" />
        <add sourceName="Exception Handling configuration"
            name="exceptionHandling" />
      </redirectSections>
    </enterpriseLibrary.ConfigurationSource>
  </configuration>
```

The Logging Block

You should log using as many listeners as you can. Why? Because it will help to ensure you have at least one log to check. One thing is certain, don't use just a single listener: file access could be blocked, there could be a database connection problem, an email server could be down, or there may be some other problem depending on the listener you are using.

I created five categories to use for logging: **Critical**, **Error**, **Warning**, **Information**, and **Verbose**. These match the values from **System.Diagnostics.TraceEventType** enumeration.

For the email listener, I use a message formatter that only includes the logged message. This listener just enables me to check for logged messages in my email while on the go, and to be notified as soon as possible of an issue through email synced to my smartphone. If I need to see information about the issue in more detail, then I check one of the other listeners. All the other listeners use the default message formatter, which includes all the information I need.

I use the Email Trace Listener for messages in the **Critical** category and for the special categories "Logging Errors & Warnings" (which are errors and warnings produced by Enterprise Library code) and the "Unprocessed Category."

I prefer to use the Database Trace Listener to get interesting indicators such as the number of errors in the last 24 hours, last week, last month, or even graphics showing a timeline of how that number changed over time. You can write any query you want against the Logging Database and retrieve any information that you think is important. I use this listener for every category except for **Verbose**. Typically, I don't use this listener to check the detailed information for a specific log message, mainly because the message is not formatted, and if you use copy and paste to get it into a text editor it appears as a single line.

For tracing to files, I decided to use three different files, and to use the Rolling Flat File Trace Listener that enables me to configure every aspect of how new files are created. My preference is for a roll interval of one day and a roll size of 5120 KB. This means that the block creates a new file every day, and if a log file size exceeds 5120 KB, the block creates another file for the same day, appending an incremental number to the file name.

When logging to files, I would recommend that you set the Message Header and Message Footer properties to indicate the start and end of each log message with words and symbols. Don't use the same text for both start and end; use values such as:

- "--- Message START -------------------------"
- "--- Message END ----------------------------"

I use one trace file as the main trace for the **Critical**, **Error**, and **Warning** categories, one trace file for **Verbose** category, and one trace file for the special category "Logging Errors & Warnings."

The final trace listener I use is the Event Log Trace Listener, which I think is the most practical to check if I need to look at a message in detail. Using the Event Viewer, I can see that each log entry has an icon that makes it very easy to locate, and the message is formatted and readable. I can also create a custom view of the events to display only the events from my application.

I use the Event Log Trace Listener for every category except for **Information** and **Verbose**.

My **Logging** assembly contains a **Logger** class with a reference to the **LogWriter** class in the Enterprise Library Logging Application Block. This **Logger** class has different methods to write messages using the Logging Application Block, but they all use the **LogWriter** class so I write all my log messages using a consistent format.

Here's an example of a simple implementation of the Logger class:

```
public class Logger
{
  private LogWriter LogWriter;
  ...
  public void Write(string message, TraceEventType traceEventType)
  {
    message += Environment.NewLine + Environment.NewLine
        + "Additional Info:" + Environment.NewLine + Environment.NewLine
        + "MachineName: " + Environment.MachineName + Environment.NewLine
        + "TimeStamp: " + DateTime.Now.ToString("dd/MM/yyyy HH:mm:ss tt");

    this.LogWriter.Write(message, traceEventType.ToString(),
                         priority: 0, eventId: 10, severity: traceEventType);
  }
}
```

The Exception Handling Block

Given the layered architecture of my application, I used this diagram from the Exception Handling Application Block as guide to implementing the exception handling in my application:

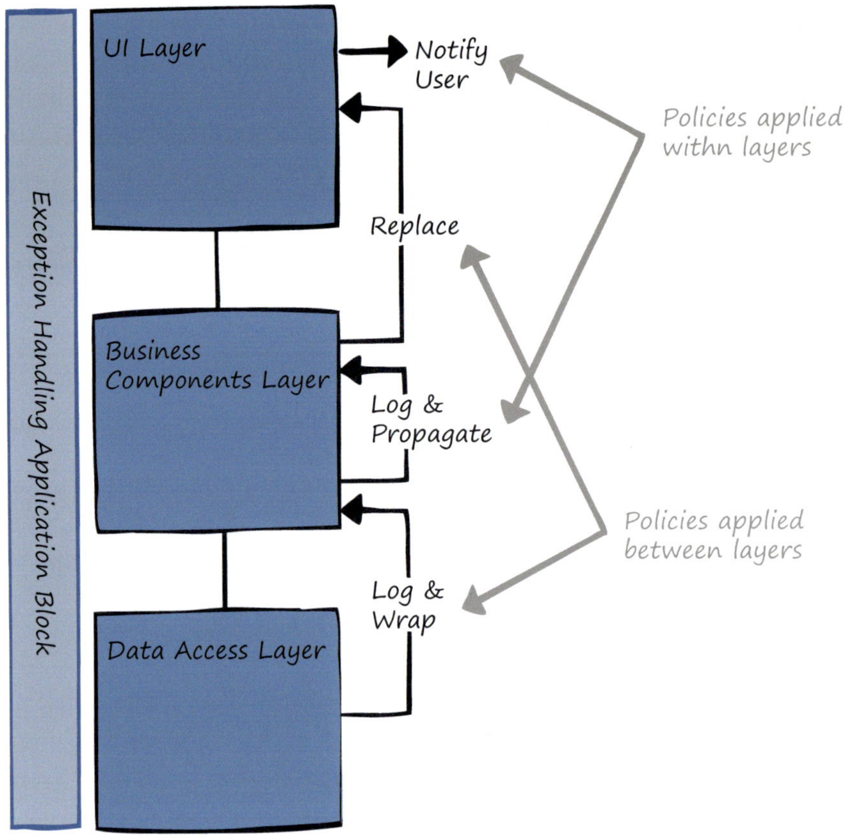

I created one exception handling policy per layer, I named the polices; Data, Business, Service, and UI. I also created a policy for unhandled exceptions at the UI Layer to use, for example, in the **Application_Error** method in our Global.asax implementation.

The architecture also includes a crosscutting layer, and in order to keep things simple, any layer that uses the crosscutting layer is responsible for handing exceptions from the crosscutting layer. This enables me to reuse the crosscutting code in projects that don't use my exception handling implementation.

The Data Layer Policy and UI Layer Unhandled Policy use a **Critical** severity and logging category, every other policy uses an **Error** severity and logging category.

I use five different types of exception to help me identify in which layer they are handled, they are: **DataException**, **BusinessException**, **ServiceException**, **UIException** and **ValidationException**.

Every logging handler uses a different Event ID to differentiate which policy is in use and at in which layer the exception occurred.

My exception handling policies log all types of exception except for **ValidationException** exceptions, which propagate all the way to the top without using an Exception Handling Application Block handler.

This list describes the behavior of each policy:

- **Data Layer Policy:** wraps the exception with a **DataException**, unless it's already a **DataException** created by the application.
- **Business Layer Policy:** replaces the exception with a **BusinessException**, unless it's already a **BusinessException** created by the application.
- **Service Layer Policy:** replaces the exception with a **ServiceException**, unless it's already a **ServiceException** created by the application or a **BusinessException** (created by the application or by the block).
- **UI Layer Policy:** replaces the exception with a **UIException**, unless it's already a **UIException** created by the application or a **ServiceException** or **BusinessException** (created by the application or by the block).
- **UI Unhandled Policy:** replaces any exception with a **UIException**.

I handle most exceptions in the public methods; I handle a small number of exceptions relating to edge cases in private and internal methods. I handle the exceptions in public methods because these are the ones used by the boundary layer.

As mentioned before, my code includes a number of custom exception definitions. It also includes an **ExceptionPolicy** enumeration to match the policy names to help make the code more maintainable and robust, and a custom **ExceptionHandler** class, which references the Enterprise Library **ExceptionManager** class, to handle application exceptions.

The **ExceptionHandler** class has only two public methods **HandlerThrow** and **HandleContinue**, and one private method used by the two public methods that adds important information into the **Data** collection of the exception.

HandleThrow lets the **ExceptionManager** instance handle the exception and throw if necessary, or it re-throws the original exception if indicated by the block.

HandleContinue uses an overload of the **HandleException** method of the **ExceptionManager** class that takes an out parameter of type **Exception** so the block does not perform a throw. The **HandleContinue** method then returns the exception indicated by the block to enable execution to continue. This is typically used at the UI Layer. For example, in an ASP.NET MVC web application I have a try/catch block inside the **Action** method of a **Controller** class. Inside the catch block, I use the **HandleContinue** method to handle the exception, capture the message from the returned exception, and add it to a list of errors in the **Model** class to show to the user.

Here's a simple implementation of the **ExceptionHandler** class:

```
public class ExceptionHandler
{
  private ExceptionManager ExceptionManager;
  ...
  public void HandleThrow(Exception ex, ExceptionPolicies policy)
  {
    MethodBase invokerMethod = (new StackTrace()).GetFrame(1).GetMethod();
    this.IncludeAdditionalInfo(ex, invokerMethod);

    bool rethrow = this.ExceptionManager.HandleException(ex, policy.ToString());

    if (rethrow)
    {
      throw ex;
    }
  }
```

```
  public Exception HandleContinue(Exception ex, ExceptionPolicies policy)
  {
    MethodBase invokerMethod = (new StackTrace()).GetFrame(1).GetMethod();
    this.IncludeAdditionalInfo(ex, invokerMethod);

    Exception lastHandlerException = null;

    bool rethrow = this.ExceptionManager.HandleException(ex,
                                                  policy.ToString(), out lastHandlerException);

    if (rethrow == true)
    {
      if (lastHandlerException == null)
      {
        return ex;
      }
      else
      {
        return lastHandlerException;
      }
    }
    else
    {
      return null;
    }
  }

  private void IncludeAdditionalInfo(Exception ex, MethodBase invokerMethod)
  {
    ex.Data["Executing Assembly"] =
       invokerMethod.Module.Assembly.FullName;
    ex.Data["Module"] = invokerMethod.Module.Name;
    ex.Data["Class"] = invokerMethod.DeclaringType;
    ex.Data["Method"] = invokerMethod.Name;
  }
}
```

Final Recommendations

Always use the Enterprise Library Configuration Tool to edit your Enterprise Library configuration files: it is very easy to use and understand, and displays the configuration information graphically. For example, if you are configuring the Logging Block you can click on a Trace Listener and instantly see which categories and which Text Formatter it is using, or you can click on a Category and see all the Trace Listeners it uses as a diagram using connector lines. It also displays help text for each element when you hover the mouse over them.

I hope that you now have a better idea of how to start using the Logging and Exception Handling Application Blocks from the Enterprise Library, if you've got any questions or just want to keep in touch, follow me at @ fabifernandez23 on twitter.

Appendix A

Enterprise Library Configuration Scenarios

The comprehensive configuration capabilities of Enterprise Library—the result of the extensible configuration system and the configuration tools it includes—make Enterprise Library highly flexible and easy to use. The combination of these features allows you to:

- Read configuration information from a wide range of sources.
- Enforce common configuration settings across multiple applications.
- Share configuration settings between applications.
- Specify a core set of configuration settings that applications can inherit.
- Merge configuration settings that are stored in a shared location.
- Create different configurations for different deployment environments.

This appendix provides an overview of the scenarios for using these features and demonstrates how you can apply them in your own applications and environments. For more information on the scenarios presented here, see "*Configuring Enterprise Library*" in the Enterprise Library Reference Documentation.

About Enterprise Library Configuration

Enterprise Library configuration information is retrieved using instances of classes that implement the **IConfigurationSource** interface, and are typically known as *configuration sources*. Figure 1 shows a high-level view of the two types of information for a configuration source and the different ways that an application's configuration can be defined and applied.

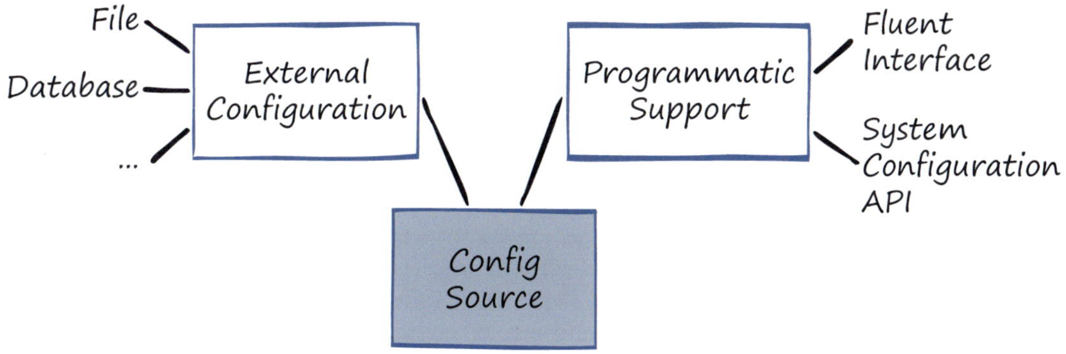

FIGURE 1
Configuration sources in Enterprise Library

External Configuration

External configuration encompasses the different ways that configuration information can reside in a persistent store and be applied to a configuration source at run time. Possible sources of persistent configuration information are files, a database, and other custom stores. Enterprise Library can load configuration information from any of these stores automatically. To store configuration in a database you can use the *SQL configuration source* that is available as a sample from the Enterprise Library community site. You can also specify one or more configuration sources to satisfy more complex configuration scenarios, and create different configurations for different run-time environments. See the section "Scenarios for Advanced Configuration" later in this appendix for more information.

Programmatic Support

Programmatic support encompasses the different ways that configuration information can be generated dynamically and applied to a configuration source at run time. Typically, in Enterprise Library this programmatic configuration takes place through the fluent interface specially designed to simplify dynamic configuration, or by using the methods exposed by the Microsoft® .NET Framework System.Configuration API.

Using the Fluent Interfaces

All of the application blocks except for the Validation Application Block and Policy Injection Application Block expose a fluent interface. This allows you to configure the block at run time using intuitive code assisted by Microsoft IntelliSense® in Visual Studio® to specify the providers and properties for the block. The following is an example of configuring an exception policy for the Exception Handling Application Block and loading this configuration into the Enterprise Library container.

```
var builder = new ConfigurationSourceBuilder();

builder.ConfigureExceptionHandling()
  .GivenPolicyWithName("MyPolicy")
  .ForExceptionType<NullReferenceException>()
    .LogToCategory("General")
      .WithSeverity(System.Diagnostics.TraceEventType.Warning)
      .UsingEventId(9000)
    .WrapWith<InvalidOperationException>()
      .UsingMessage("MyMessage")
    .ThenNotifyRethrow();

var configSource = new DictionaryConfigurationSource();
builder.UpdateConfigurationWithReplace(configSource);
ExceptionPolicy.SetExceptionManager(new ExceptionPolicyFactory(configSource).CreateManager());
```

Scenarios for Advanced Configuration

The Enterprise Library stand-alone configuration console and the Visual Studio integrated configuration editor allow you to satisfy a range of advanced configuration scenarios based on external configuration sources such as disk files. When you use the configuration tools without specifying a configuration source, they default to using the System Configuration Source to create a single configuration file that contains the entire configuration for the application. Your application will expect this to be named App.config or Web.config (depending on the technology you are using), and will read it automatically.

You can select **Add Configuration Settings** on the **Blocks** menu to display the section that contains the default system configuration source. If you click the chevron arrow to the right of the **Configuration Sources** title to open the section properties pane you can see that this is also, by default, specified as the **Selected Source**—the configuration source to which the configuration generated by the tool will be written. When an application that uses Enterprise Library reads the configuration, it uses the settings specified for the selected source unless a different source is explicitly requested.

The following sections describe the common scenarios for more advanced configuration that you can accomplish using the configuration tools. Some of these scenarios require you to add additional configuration sources to the application configuration.

Scenario 1: Using the Default Application Configuration File

This is the default and simplest scenario. You configure your application using the configuration tool without adding a **Configuration Sources** section or any configuration sources. You must specify either your application's App.config or Web.config file when you save the configuration, or use the configuration tool to edit an existing App.config or Web.config file.

Scenario 2: Using a Non-default Configuration Store

In this scenario, you want to store your configuration in a file or other type of store, instead of in the application's App.config or Web.config file. To achieve this you:

Use the configuration tools to add a suitable configuration source to the **Configuration Sources** section. If you want to use a standard format configuration file, add a file-based configuration source. To store the configuration information in a different type of store, you must install a suitable configuration source. You can use the sample SQL configuration source that is available from the Enterprise Library community site at *http://entlib.codeplex.com* to store your configuration in a database.

Set the relevant properties of the new configuration source. For example, if you are using the built-in file-based configuration source, set the **File Path** property to the path and name for the configuration file.

Set the **Selected Source** property in the properties pane for the **Configuration Sources** section to your new configuration source. This updates the application's default App.config or Web.config file to instruct Enterprise Library to use this as its configuration source.

Scenario 3: Sharing the Same Configuration between Multiple Applications

In this scenario, you want to share configuration settings between multiple applications or application layers that run in different locations, such as on different computers. To achieve this, you simply implement the same configuration as described in the previous scenario, locating the configuration file or store in a central location. Then specify this file or configuration store in the settings for the configuration source (such as the built-in file-based configuration source) for each application.

Scenario 4: Sharing Configuration Sections across Multiple Applications

In this scenario, you have multiple applications or application layers that must use the same shared configuration for some application blocks (or for some sections of the configuration such as connection strings). Effectively, you want to be able to redirect Enterprise Library to some shared configuration sections, rather than sharing the complete application configuration. For example, you may want to specify the settings for the Logging Application Block and share these settings between several applications, while allowing each application to use its own local settings for the Exception Handling Application Block. You achieve this by redirecting specific configuration sections to matching sections of a configuration store in a shared location. The steps to implement this scenario are as follows:

Use the configuration tools to add a suitable configuration source for your application to the **Configuration Sources** section. This configuration source should point to the shared configuration store. If you want to use a standard format configuration file as the shared configuration store, add a file-based configuration source. To store the shared configuration information in a different type of store, you must install a suitable configuration source. You can use the sample SQL configuration source that is available from the Enterprise Library community site at *http://entlib.codeplex.com* to store your configuration in a database.

Set the relevant properties of the shared configuration source. For example, if you are using the built-in file-based configuration source, set the **File Path** property to the path and name for the application's configuration file.

1. Set the **Selected Source** property in the properties pane for the **Configuration Sources** section to **System Configuration Source**.
2. Click the plus-sign icon in the **Redirected Sections** column and click **Add Redirected Section**. A redirected section defines one specific section of the local application's configuration that you want to redirect to the shared configuration source so that it loads the configuration information defined there. Any local configuration settings for this section are ignored.
3. In the new redirected section, select the configuration section you want to load from the shared configuration store using the drop-down list in the **Name** property. The name of the section changes to reflect your choice.
4. Set the **Configuration Source** property of the redirected section by selecting the shared configuration source you defined in your configuration. This configuration source will provide the settings for the configuration sections that are redirected.
5. Repeat steps 4, 5, and 6 if you want to redirect other configuration sections to the shared configuration store. Configuration information for all sections for which you do not define a redirected section will come from the local configuration source.
6. To edit the contents of the shared configuration store, you must open that configuration in the configuration tools or in a text editor; you cannot edit the configuration of shared sections when you have the local application's configuration open in the configuration tool. If you open the shared configuration in the configuration tool, ensure that the **Selected Source** property of that configuration is set to use the system configuration source.

   *You cannot share the contents of the **Application Settings** section. This section in the configuration tool stores information in the standard **<appSettings>** section of the configuration file, which cannot be redirected.*

Scenario 5: Applying a Common Configuration Structure for Applications

In this scenario you have a number of applications or application layers that use the same configuration structure, and you want to inherit that structure but be able to modify or add individual configuration settings by defining them in your local configuration file. You can specify a configuration that inherits settings from a parent configuration source in a shared location, and optionally override local settings. For example, you can configure additional providers for an application block whose base configuration is defined in the parent configuration. The steps to implement this scenario are as follows:

Use the configuration tools to add a suitable configuration source for your application to the **Configuration Sources** section. This configuration source should point to the shared configuration store. If you want to use a standard format configuration file as the shared configuration store, add a file-based configuration source. To store the shared configuration information in a different type of store, you must install a suitable configuration source. You can use the sample SQL configuration source that is available from the Enterprise Library community site at *http://entlib.codeplex.com* to store your configuration in a database.

Set the relevant properties of the shared configuration source. For example, if you are using the built-in file-based configuration source, set the **File Path** property to the path and name for the application's configuration file.

Set the **Parent Source** property in the properties pane for the **Configuration Sources** section to your shared configuration source. Leave the **Selected Source** property in the properties pane set to **System Configuration Source**.

Configure your application in the usual way. You will not be able to see the settings inherited from the shared configuration source you specified as the parent source. However, these settings will be inherited by your local configuration unless you override them by configuring them in the local configuration. Where a setting is specified in both the parent source and the local configuration, the local configuration setting will apply.

To edit the contents of the shared parent configuration store, you must open that configuration in the configuration tools or in a text editor; you cannot edit the configuration of parent sections when you have the local application's configuration open in the configuration tool. If you open the parent configuration in the configuration tool, ensure that the **Selected Source** property of that configuration is set to use the system configuration source.

The way that the configuration settings are merged, and the ordering of items in the resulting configuration, follows a predefined set of rules. These are described in detail in the *Enterprise Library Reference Documentation*.

Scenario 6: Managing Configuration in Different Deployment Environments

In this scenario, you want to be able to define different configuration settings for the same application that will be appropriate when it is deployed to different environments, such as a test and a production environment. In most cases the differences are minor, and usually involve settings such as database connection strings or the use of a different provider for a block. Enterprise Library implements this capability using a feature called environmental overrides. The principle is that you specify override values for settings that are different in two or more environments, and the differences are saved as separate delta configuration files. The administrator then applies these differences to the main configuration file when the application is deployed in each environment. To achieve this:

1. Use the configuration tools to add a file-based configuration source to the **Configuration Sources** section.
2. Open the **Environments** menu and click **New Environment** for each desired environment. This adds a drop-down list for each environment, **Overrides on *Environment Name***, to each configuration element. If you select **Override Properties** in an environment's drop-down list, you can specify the setting overrides for the element in that environment.
3. Open the properties pane for each of the environments you added to your configuration by clicking the chevron arrow to the right of the environment title, and set the **Environment Delta File** property to the path and name for the delta file for that environment.
4. Specify the overrides to the configuration elements that need environment-specific settings.
5. Save the configuration. The configuration tool generates a normal (.config) file and a delta (.dconfig) file for each environment. The delta file(s) can be managed by administrators, and stored in a separate secure location, if required. This may be appropriate when, for example, the production environment settings should not be visible to developers or testers.

6. To create a run-time merged configuration file (typically, this is done by an administrator):
   - Open the local configuration (.config) file.
   - Select **Open Delta File** from the **Environments** menu and load the appropriate override configuration (.dconfig) file.
   - Set the **Environment Configuration File** property in the properties pane for the environment to the path and name for the merged configuration file for that environment.
   - Right-click on the title of the environment and click **Export Merged Environment Configuration File**.
7. Deploy the merged configuration file in the target environment.

Enterprise Library also contains a command-line utility named MergeConfiguration.exe that you can use to merge configuration and delta files if you do not have the configuration console deployed on your administrator system. It can also be used if you wish to automate the configuration merge as part of your deployment process. For more information about MergeConfiguration.exe, see "*Configuring a Deployment Environment*" in the Enterprise Library Reference Documentation.

You cannot use environmental overrides with redirected sections or inherited configuration settings. You can only use them when the entire configuration of your application is defined within a local configuration source.

For more information on all of the scenarios presented here, see "*Configuring Enterprise Library*" in the Enterprise Library Reference Documentation.

More Information

All links in this book are accessible from the book's online bibliography on MSDN at *http://aka.ms/el6biblio*.

For more information on the scenarios presented here, see "*Configuring Enterprise Library*" in the Enterprise Library Reference Documentation.

To store configuration in a database you can use the *SQL configuration source* that is available as a sample from the Enterprise Library community site.

The way that the configuration settings are merged, and the ordering of items in the resulting configuration, follows a predefined set of rules. These are described in detail in the *Enterprise Library Reference Documentation*.

For more information about MergeConfiguration.exe, see "*Configuring a Deployment Environment*" in the Enterprise Library Reference Documentation.

General Links:

- There are resources to help if you're getting started with Enterprise Library, and there's help for existing users as well (such as the breaking changes and migration information for previous versions) available on the Enterprise Library Community Site at *http://www.codeplex.com/entlib/*.
- For more details about the Enterprise Library Application Blocks, see the *Enterprise Library Reference Documentation* and the *Enterprise Library 6 Class Library*.
- You can download and work through the Hands-On Labs for Enterprise Library, which are available at *http://aka.ms/el6hols*.
- You can download the Enterprise Library binaries, source code, Reference Implementation, or QuickStarts from the Microsoft Download Center at *http://go.microsoft.com/fwlink/p/?LinkID=290898*.
- To help you understand how you can use Enterprise Library application blocks, we provide a series of simple example applications that you can run and examine. To obtain the example applications, go to *http://go.microsoft.com/fwlink/p/?LinkID=304210*.

Appendix B Encrypting Configuration Files

Enterprise Library supports encryption of configuration information. Unless your server is fully protected from both physical incursion and remote incursion over the network, you should consider encrypting any configuration files that contain sensitive information, such as database connection strings, passwords and user names, or validation rules.

You can select any of the encryption providers that are included in your system's Machine.config file. Typically, these are the **DataProtectionConfigurationProvider**, which uses the Windows Data Protection API (DPAPI), and the **RsaProtectedConfigurationProvider**, which uses RSA encryption. The settings for these providers, such as where keys are stored, are also in the Machine.config file. You cannot edit this file with a configuration tool; instead, you must modify it using a text editor or an operating system configuration tool. You can also define and use new providers with different settings in your application's configuration file.

If you deploy your application to Windows Azure, you should also carefully consider how to encrypt configuration settings stored in Windows Azure. One approach to consider is using the *"Pkcs12 Protected Configuration Provider."*

As an example of the effect of this option, the following is a simple unencrypted configuration for the Data Access block.

```
<dataConfiguration defaultDatabase="Connection String" />
<connectionStrings>
 <add name="Connection String"
      connectionString="Database=TheImportantOne; Server=WEHAVELIFTOFF;
                        User ID=secret; Password=DontTellNE1"
      providerName="System.Data.SqlClient" />
</connectionStrings>
```

When you specify the **DataProtectionConfigurationProvider** option, the resulting configuration section looks like the following.

```
<dataConfiguration configProtectionProvider="DataProtectionConfigurationProvider">
 <EncryptedData>
  <CipherData>
   <CipherValue>AQAAANCMnd8BFdERjHoAwE/Cl+sBAAAAc8HVTgvQB0quQI81ya0uH
     yTmSDdYQNdiSohA5Fo6bWOqhOR5V0uxdcfNUgKhUhuIAhl5RZ8W5WD8M2CdMiqG
     ...
     JyEadytIBvTCbmvXefuN5MWT/T
   </CipherValue>
  </CipherData>
 </EncryptedData>
</dataConfiguration>
```

```
<connectionStrings configProtectionProvider="DataProtectionConfigurationProvider">
 <EncryptedData>
  <CipherData>
   <CipherValue>AQAAANCMnd8BFdERjHoAwE/Cl+sBAAAAc8HVTgvQB0quQI81ya0uH
      ...
      zBJp7SQXVsAs=</CipherValue>
  </CipherData>
 </EncryptedData>
</connectionStrings>
```

If you only intend to deploy the encrypted configuration file to the server where you encrypted the file, you can use the **DataProtectionConfigurationProvider**. However, if you want to deploy the encrypted configuration file on a different server, or on multiple servers in a Web farm, you should use the **RsaProtectedConfigurationProvider**. You will need to export the RSA private key that is required to decrypt the data. You can then deploy the configuration file and the exported key to the target servers, and re-import the keys. For more information, see "*Importing and Exporting Protected Configuration RSA Key Containers*."

Of course, the next obvious question is "How do I decrypt the configuration?" Thankfully, you don't need to. You can open an encrypted file in the configuration tools as long as it was created on that machine or you have imported the RSA key file. In addition, Enterprise Library blocks will be able to decrypt and read the configuration automatically, providing that the same conditions apply.

More Information

All links in this book are accessible from the book's online bibliography on MSDN at *http://aka.ms/el6biblio*.

If you deploy your application to Windows Azure, you should also carefully consider how to encrypt configuration settings stored in Windows Azure. One approach to consider is using the "*Pkcs12 Protected Configuration Provider*."

For more information on exporting the RSA private key that is required to decrypt the data, see "*Importing and Exporting Protected Configuration RSA Key Containers*."

General Links:

- There are resources to help if you're getting started with Enterprise Library, and there's help for existing users as well (such as the breaking changes and migration information for previous versions) available on the Enterprise Library Community Site at *http://www.codeplex.com/entlib/*.
- For more details about the Enterprise Library Application Blocks, see the *Enterprise Library Reference Documentation* and the *Enterprise Library 6 Class Library*.
- You can download and work through the Hands-On Labs for Enterprise Library, which are available at *http://aka.ms/el6hols*.
- You can download the Enterprise Library binaries, source code, Reference Implementation, or QuickStarts from the Microsoft Download Center at *http://go.microsoft.com/fwlink/p/?LinkID=290898*.
- To help you understand how you can use Enterprise Library application blocks, we provide a series of simple example applications that you can run and examine. To obtain the example applications, go to *http://go.microsoft.com/fwlink/p/?LinkID=304210*.

Index

B

C

D

G

H

I

J

K

L

M

N

O

P

U

V

W

X

Colophon

The origami on the opposite page and on the cover of the Enterprise Library Developer's Guide is a net of a ***regular octahedron***. This 8-faced polyhedron represents the 8 integrated application blocks of Enterprise Library while the origami represents the philosophy of the patterns & practices team providing guidance for you to build complex systems.

According to math.wikia.com:

> *A regular octahedron is a Platonic solid composed of eight equilateral triangles, four of which meet at each vertex. The aesthetic beauty and symmetry of octahedron has made it one of the favorite subjects of geometers for thousands of years. The octahedron is unique among the Platonic solids in having an even number of faces meeting at each vertex. Consequently, it is the only member of that group to possess mirror planes that do not pass through any of the faces. In the physical world, natural crystals of diamond, alum, or fluorite are commonly octahedral.*

There are 11 distinct nets for the octahedron. This symbolizes various possible combinations of Enterprise Library application blocks.

An octahedron can be placed with its center at the origin and its vertices on the coordinate axes; the Cartesian coordinates of the vertices are:

$(\pm1, 0, 0)$;

$(0, \pm1, 0)$;

$(0, 0, \pm1)$.

Octahedra themselves can further be used as building blocks to construct other fascinating polyhedrons.

To download and print the octahedron origami, go to *http://aka.ms/EL6origami*

For more info, see:

- Weisstein, Eric W., "*Octahedron*" from MathWorld
- *http://math.wikia.com/wiki/Octahedron*
- *http://math.wikia.com/wiki/Platonic_solid*

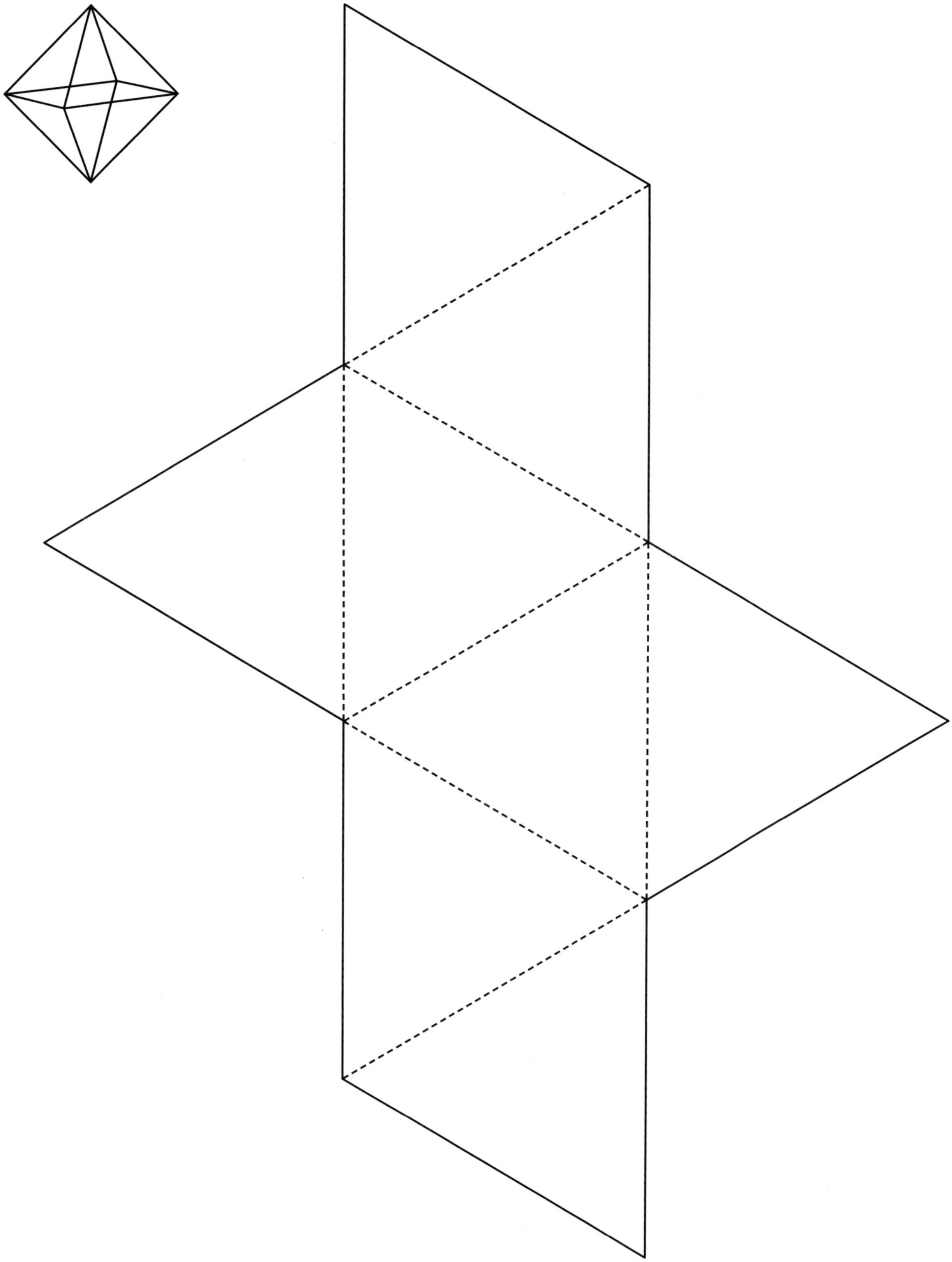

Made in the USA
Las Vegas, NV
03 November 2026

57424841R00148